Natural Science
in Western History

Natural Science in Western History

VOLUME I

From Ancient Times to Newton

Frederick Gregory
UNIVERSITY OF FLORIDA

HOUGHTON MIFFLIN COMPANY Boston   New York

Publisher: Suzanne Jeans
Senior Sponsoring Editor: Nancy Blaine
Senior Marketing Manager: Katherine Bates
Senior Development Editor: Lisa Kalner Williams
Senior Project Editor: Christina Horn
Art and Design Manager: Jill Haber
Cover Design Manager: Anne S. Katzeff
Senior Photo Editor: Jennifer Meyer Dare
Composition Buyer: Chuck Dutton
New Title Project Manager: Susan Peltier
Editorial Associate: Adrienne Zicht
Marketing Assistant: Lauren Bussard
Editorial Assistant: Carrie Parker

Cover image: Joos van Ghent, portrait of Claudius Ptolemy, Greek astronomer. © Erich Lessing/Art Resource, NY.

Printed in the U.S.A.

Library of Congress Control Number: 2007933468

ISBN-13: 978-0-618-22411-1
ISBN-10: 0-618-22411-4

123456789-VHO-11 10 09 08 07

◎

*for Kalée and Laura
and their wonderful families*

Brief Contents

Contents

Preface

Historians of science face the same challenges that confront any author who attempts to render an account of former times. As historians, we do our best to represent past events and achievements in the way that people experienced them rather than as we might imagine them. In the ancient world, for example, there was no such thing as "science" in anything like the twenty-first century meaning of the word. Although today we distinguish science from technology and science from religion, no clean separation of these endeavors existed in antiquity. To use present-day conceptions of natural science and other disciplines in our treatment of the past would surely distort the motivations, goals, and the practices of earlier peoples.

Having said that, we should not imagine that we can put ourselves wholly into the shoes of Babylonians or Greeks who lived so long ago, or even, for that matter, of those who lived in the last century. Try as we might not to bring the perspective of our own times to our investigation of the past, we will unwittingly smuggle some assumptions and perspectives of the culture in which we live into our portrait. When early in the twentieth century George Sarton, the founder of the discipline of history of science in the United States, chose the Egyptian goddess Isis to represent the challenge humans have faced in trying to discover how nature works, he did so because Isis said of herself that she was everything that ever existed and that "no mortal had ever disclosed her robe." His choice of Isis to stand for natural science was consistent with a long-standing Western tradition that has regarded nature as female and natural science as the process of uncovering her secrets. It reveals, as historian of science Carolyn Merchant has made clear, that our modern understanding of how to represent not only the ancient world, but also that of natural science itself, has been subtly informed by the sexual culture of the West.

How are we to handle a situation in which what we strive to do—to represent the past as it was experienced by those who lived it—must be acknowledged as impossible from the start? The first answer is that we should not conclude that nothing from the past is accessible to us. Just because we cannot give a completely unbiased account does not mean that we cannot do a reasonably decent job of representing the past on its own terms. In any event, borrowing from a common wish of the scientist to find the truth of nature, I adopt here the following procedure as an operating principle: I will aim at reaching the truth, even though I acknowledge that I cannot hope to achieve it. In claiming this maxim I shall not hesitate to make judgments about historical evidence at the same time that I register my hope to remain sensitive to my own fallibility.

All this means that what follows is a personal interpretation of the history of science that reflects the interests, abilities, and biases of the author. The title itself, *Natural Science in Western History,* already announces that it does not include developments in the East, nor does it include technology or the social sciences. As necessary and important as these subjects are, they exceed the scope of this endeavor. Every reader will no doubt discover that a topic he or she considers essential is not mentioned, while other matters regarded as spurious receive too much attention. Further, readers will have preferences that vary with the choices found here about how much the focus should be on the practice of science and on its social and cultural context as opposed to the scientific ideas themselves.

Undertaking a history of science textbook from antiquity to the present is a daunting task indeed. My introduction to the idea occurred in the 1980s, when the Department of History at the University of Florida sponsored a workshop that brought to Gainesville five historians of science—David Lindberg of the University of Wisconsin, William Ashworth of the Linda Hall Library in Kansas City, Sharon Kingsland of Johns Hopkins University, Thomas Hankins of the University of Washington, and the late Frederic Holms of Yale University—to explore the prospect of taking on such a project. That effort bore one important concrete product: David Lindberg's wonderful text on ancient and medieval science, an invaluable resource on which I have relied in this survey. Although Lindberg fulfilled his part of the projected whole, none of the rest of us did, and the collective endeavor foundered. David continued, however, to encourage me, even to the point of reading an early draft of the first chapter I attempted, on Newton. I owe him much.

Five or six years ago Nancy Blaine of Houghton Mifflin, who was visiting the University of Florida, asked during a conversation in my office why history of science was lacking a recent textbook. When I mentioned my previous flirtation with the idea of writing such a work, she persuaded me to submit a proposal. What interested her was my characterization of the type of text I would write, one that at the time was different from works in the field and one that, I am pleased to say, remains different from the textbooks that have since appeared. My text would be aimed at the average undergraduate and would attempt to do for history of science what U.S. history and Western Civilization texts have done for their respective fields. Having taught Western Civ for many years myself, I knew the value of an essentially chronological narrative of the subject matter, grouped under headings and subheadings, whose language and style were specifically designed to be accessible to university students. And so I began.

Acknowledgments

When a project lasts over several years, the number of people to thank increases rapidly. Locally, the Department of History has understood that a textbook project also serves the larger profession, as has the College of Liberal Arts and Sciences at the University of Florida. I have already made clear the great debt owed to Nancy Blaine of Houghton Mifflin. Not only did she provide the impetus to begin but she has also remained involved, always with words of encouragement and helpful advice. Numerous editors in various capacities at Houghton Mifflin have played constructive roles, including Stan Yarbro, Julie Swasey, Katie White, Adrienne Zicht, Christina Horn, Marilyn Rothenberger, Peggy Flanagan, and Bruce Carson.

To the many referees of the chapters, although unknown to me but listed by the publisher below, I owe a great deal. They have corrected errors and offered critiques on every aspect of the work. Many times I have heeded their advice, and at other times I have disagreed with their suggestions, but always I have appreciated and respected their views. The following individuals served as reviewers of the manuscript: David Boersema, Pacific University; Dan Burton, University of North Alabama; Jack Holl, Kansas State University; Florence Hsia, University of Wisconsin–Madison; Bruce Hunt, University of Texas; Michel Janssen, University of Minnesota; John Kemp, Truckee Meadows Community College; Dr. William C. Kimler, North Carolina State University; Kristie Macrakis, Michigan State University; Elizabeth Green Mussleman, Southwestern University (Texas); Timothy Moy, University of New Mexico; Warren Neal, Oklahoma City Community College; Anna Marie Roos, University of Minnesota; Andrea Rusnock, University of Rhode Island; David Sepkoski, Oberlin College; Warren Van Egmond, Arizona State University; Alice Walters, University of Massachusetts, Lowell; and G. J. Weisel, Penn State Altoona.

This project has demanded much more from my wife Patricia than does my regular scholarly research. Not only was it demanding and all consuming of my time and attention over a very long time, but it also produced far more frustration than does usual archival work. Organizing the material was by far the greatest challenge, eliciting constant complaints. I finally realized how this was impacting Tricia when she said to me one day, "You can only say how hard it is to organize your material *two more times,* so choose them wisely." As always, however, she has been there for me, ready to put things into perspective by taking me dancing.

Gainesville, Florida
July 2007

CHAPTER 1

───────◎───────

The Ancient Western Heritage

In the world of the ancient Near East there was no such thing as natural science as we understand it today, in the twenty-first century. The early civilizations of Egypt and Mesopotamia did, however, attempt to make sense of nature. They achieved a coherence in their understanding of nature through the development of a tradition of myths, and sometimes with the help of mathematical calculation. Natural processes, as we shall see, were directly linked to the presence of gods.

In ancient Greece, prior to the time of Plato, depictions of the natural world also revolved around the activities of divine beings. Then, at the end of the sixth century B.C., a distinctly Greek approach that emphasized the world's *rational* structure made its appearance. Scholars have referred to the unique perspective on nature that began to emerge in the Ionian region at this time as the Greek miracle. This ancient Ionian view of nature contained aspects that contrasted with the more personal understandings of the natural world that had been dominant previously in the West.

Two thinkers from ancient Greece whose ideas proved to be crucial for the development of systematic depictions of the natural world were the philosophers Plato (ca. 427–ca. 347 B.C.) and Aristotle (384–322 B.C.). As we shall see, they were responsible for defining fundamental issues that would occupy thinkers for centuries to come.

◎ Nature in Ancient Civilizations ◎

The Near East

In what one scholar has identified as a mythopoeic view of nature, ancient peoples of Egypt and Mesopotamia viewed nature itself as an expression of deities, and natural processes as the direct actions of these gods. For example, ancient Egyptians invoked the sky goddess, Nut, to account for the appearance of the heavens. As she

The Sky Goddess, Nut

bent over the Earth, touching it with her toes and fingers, her body was speckled with the stars. She was also the mother of the sun, which she swallowed every evening and gave birth to again each morning.

These mythological descriptions of nature did not prevent the Egyptians from devising means by which to use their knowledge of the movements of the heavens for religious, agricultural, and calendrical purposes. They marked the end of night, for example, not only by Nut giving birth to the sun, but also by the appearance of a pattern among the many stars that rose just before sunrise and marked where the sun would come up.

Over time they noticed, however, that the star pattern was appearing more and more in advance of the sun, so that after ten days it was rising in total darkness and a different pattern had to be chosen to mark the last hour of the night. As the process repeated, additional marker patterns were visible above the horizon in the darkness just before sunrise. They accumulated within a band, later called the band of the zodiac, which ran across the sky at the angle that the sun itself traversed once it rose and traveled across the sky.

The effect of these motions was that the sun, in lagging behind the stars, appeared to move through an entire series of marker patterns. (They were also aware that the moon lagged behind the stars, at a much greater rate than did the sun.) After using thirty-six different patterns, each lasting ten days, to denote the end of night, the Egyptians observed that the thirty-seventh served for only five days before the first pattern reappeared and began the series all over again. This clearly meant that a cycle of some sort was traversed in $(10 \times 36) + 5 = 365$ days. On the Egyptian calendar, which was established well before the Middle Kingdom (2040–1640 B.C.), the year took 360 plus 5 extra so-called epagomenal days that were observed at the end.

The Egyptians grouped three of the marker patterns together to form a month, so their year was twelve months plus five epagomenal days. We might expect that they would have divided the day and night into eighteen divisions each since it took thirty-six star patterns to go all around the sky. But they divided them unevenly into twenty-four periods—the night into twelve and the day into ten, with two more periods for the twilight of morning and evening. This meant that periods themselves lasted longer or shorter, depending on the seasonal fluctuation in the amount of daylight time. The Greeks later took over this division of the day-night cycle into twenty-four hours, making all the hours of equal length.

The ancient Babylonians devised a sexagesimal number system. Unlike the decimal system, which is based on 10, the Babylonians used 60 as a base. By repeating and spacing numerals, designated by a wedge-shaped mark, they were able to express very large numbers. Thus ▼▼▼ was 3, while ▼   ▼▼▼ was 63 $(1 \times 60 + 3)$ and, similarly, ▼   ▼   ▼▼▼ was 3663 $(1 \times 60^2 + 1 \times 60 + 3)$. From ancient tablets it is also clear that the Babylonians expressed fractions. To do this the spacings were interpreted from the context to indicate *division* by powers of 60 moving to the right instead of multiples of powers of 60 moving to the left. Hence, depending on the context, ▼   ▼▼▼ could also mean $63/3600$ $(1/60 + 3/60^2)$. There are even tablets that indicate solutions to complex mathematical problems; for example, on one tablet the side of a square is determined through a series of steps that are equivalent to the solution of a quadratic equation.

By 400 B.C. the Babylonian astronomers were using twelve of their own marker patterns of stars to mark off the year. Each constellation took up 30° of the total of 360° that ran completely around the zodiac. Babylonian priests carefully prepared charts that recorded a great deal of astronomical data. They noted when eclipses occurred and they noted the positions of the sun and the moon in the zodiac over the course of the year. Within this huge amount of data they identified repeating numerical patterns that enabled them to predict important celestial events, such as the appearance of the new moon or even a lunar eclipse.

The priests also observed that the sun and moon were not the only objects in the zodiacal band that moved against the background of the remaining stars from night to night. There were five other stars that exhibited their own proper motion, although the motion was not confined to the "lagging behind" kind of motion the sun and moon underwent. These five stars—which we know as the planets Mercury, Venus, Mars, Jupiter, and Saturn—displayed various motions, sometimes lagging behind the rest of the stars and sometimes turning around and moving ahead of them. Not surprisingly, the priests associated these wandering stars with the gods, whose actions were responsible for these unusual motions.

Early Greek Conceptions of the Natural World

Some of these Egyptian and Babylonian conceptions made their way to the Greek world, which by the fifth century B.C. comprised the territory surrounding the Aegean Sea. The Greeks possessed their own rich mythology to explain the natural world. The sun's journey across the sky, for example, was due to the god Apollo, who daily rode his chariot across the sky. The Greek writers Homer and Hesiod had described the life of the gods in considerable detail by the end of the eighth century B.C. and the mythology they articulated continued to flourish for centuries.

The advent of Greek philosophy.   In the sixth century, however, a new mode of thought, Greek philosophy, appeared alongside and sometimes mingled with these mythological accounts. The most distinguishing feature of this philosophy was that it explained natural phenomena not through a direct appeal to the actions of the gods, but as the course of nature itself. The philosophers lessened the role of personal agency in their explanations and increased the role of impersonal agency. For Anaximander, a philosopher who flourished during the mid-sixth century B.C. in the city of Miletus on the west coast of present-day Turkey, thunder and lightning had nothing to do with Zeus and his thunderbolts; rather, they were caused by wind breaking out of clouds.

Other thinkers from Miletus also removed personal agency from the explanations of nature and natural process. These early Milesians of the sixth century embraced the central conviction that the cosmos followed regular and necessary patterns of behavior, that it was not chaotic, that human existence was part of the natural world order. They went further. They were convinced that they could use their own powers of reason to uncover these patterns.

The Milesians asked new questions. They wanted to know what things were made of, how they moved or otherwise changed over time, why they had the shape and weight they did, and why they were located where they were. One of the most basic questions that sparked their curiosity had to do with the basic stuff from which everything else was made. Various Milesians answered this question with different substances. Thales, one of the earliest, said water was the basic stuff of things, while later in the century Anaximines said it was air. At the beginning of the fifth century B.C., Heraclitus, from nearby Ephesus, identified nature's most basic constituent as fire.

In posing the question about the basic stuff from which everything else was made, these early Greek philosophers introduced a crucial assumption that would continue to mark the Western tradition from this time forward. Behind the question lay the supposition that the diversity found in nature derived from a fundamental unity. If this was so, then it was possible to give reasons for this diversity in terms of the underlying unity. It was possible, in other words, to explain nature rationally.

Pythagoras and his followers.   Another important Greek figure from the late sixth and early fifth centuries B.C. was Pythagoras, who is familiar to us from the famous mathematical theorem attributed to him. Much about the man is shrouded in mythology. According to various ancient sources he traveled to Egypt and then to Babylonia, where he became familiar with new mathematical techniques. On his return he set up a religious community among Greek colonists in southern Italy. Here, it is said, the group lived together communally and developed its own religious ritual practices of not eating meat or fish and not wearing wool or leather. The group also cultivated religious secrecy, especially about the propositions they uncovered in their mathematical school.

Pythagoras and his followers shared with the other late-sixth-century philosophers the need for a unified explanation of our experience of the senses, but they chose as their unifying entity not a material substance like water or air, but an abstract entity—*number*. An important illustration of what they meant by asserting that things were made of number was their investigation of the numerical ratios that corresponded to the harmonious pitches of musical tones. A string vibrates to produce a musical sound, as in a lyre. Shortening the vibrating portion of the string to half its length changes the pitch by an octave. Shortening it to a third, two-thirds, or three-fourths produces other pleasant harmonies.

Numerical relationships can present us with fascinating properties, some of them even bewildering. As the Pythagoreans uncovered new relationships, they invested them with greater and greater importance. That is, they became convinced that the mysterious relationships they uncovered were of divine origin. This belief separated them from other early Greek thinkers, who had completely removed divine agency from nature. The Pythagoreans embraced a religious outlook that invested the foundation of reality with number as its most basic constituent part.

Mathematical relationships are among the most rational we know. Therefore, the Pythagorean conviction that number governed nature meant that nature was describable in mathematical terms, which is to say that nature was rational. In other words, the Pythagoreans reinforced the new Greek assumption that nature was rationally explainable. They also demonstrated one manner in which strong religious belief and a firm commitment to rational explanation could go hand in hand.

When the Pythagoreans uncovered a mathematical relationship that was not rational, it challenged the very basis of their religion. From the famous theorem about the sides of a right triangle, they knew that the square built on the hypotenuse was equal in area to the sum of the squares erected on the other two sides. That led them to consider the case of an isosceles right triangle with sides equal to one unit of length. What is the length of the hypotenuse? They assumed that the answer to this question could be found, since reality was rational; that is, they assumed that they could find the common measure between the hypotenuse and one of the sides so that they could express both lengths as certain multiples of this common measure.

But, because they examined the question logically, they became convinced that there was no unit small enough that could measure the hypotenuse as an integral number of multiples of itself and at the same time measure a side with a different integral number of multiples of itself. The ratio of the side to the hypotenuse, in other words, could not be expressed as a ratio of integral numbers p/q where p and q had no common factors.

On realizing that there were some magnitudes that were not commensurate with each other it appeared to the Pythagoreans that they had discovered an instance where the world did not conform to number. Here reality was being *a logikos,* not rational. In today's mathematics the numbers assigned to such magnitudes are known as irrational numbers. In the previously mentioned triangle the irrational length of the hypotenuse is $\sqrt{2}$. But if there were such irrationalities, then perhaps even reality could not be guaranteed to be rational and to make sense.

This conclusion directly challenged the conviction of the early Greek philosophers that nature was in fact rational. If reality was ultimately chaotic, then what was the purpose of trying to explain it? Various legends cropped up about the dire consequences that followed for those who disclosed the knowledge that the hypotenuse bore an irrational ratio to the side of a right triangle.

◎ The Cosmos of Plato and Aristotle ◎

The capacity of reason to produce a contradiction was instrumental in the well-known paradoxes of the philosopher Zeno, who flourished during the middle of the fifth century B.C. Zeno was convinced that reason could be trusted, so much so that he was willing to discard obvious conclusions of experience when they differed from what reason said must be true. For example, our experience certainly tells us that things change, but Zeno trusted reason over experience and used it to argue that change was impossible.

Zeno set out to prove that change was an illusion of surface reality that we accept from our sense experience. That experience, however, does not draw directly on the basic unchanging reality accessible to reason. By using our reason we can convince ourselves that change is impossible, even the change of place so obvious to us in the experience of motion. Zeno's approach was to make an assumption based on experience, then to use reason to challenge the conclusions of experience. That, he felt, exposed the assumption of sense experience as erroneous.

Consider Achilles, who is pursuing a tortoise. We assume from observational experience that, because change of place is possible, Achilles will easily overtake the slow-moving tortoise. But the analysis of reason shows that this cannot be the case. In order for Achilles to catch the tortoise, he must first go to the point where the tortoise was when the pursuit commenced. But in that time the tortoise will have moved ahead. Now Achilles must go to the point where the tortoise is now. But when he reaches that point the tortoise will no longer be there either. It will always be so. The tortoise will always be ahead of Achilles, even if the distances between them keep shrinking. Rational analysis has shown the assumption that Achilles will pass the tortoise to be wrong. Since that assumption is based on the possibility of change, change has been shown to be impossible.

The development of Greek thought broadened in the fourth century B.C. to include a diversity of concerns, including political and ethical philosophy. The works of two men, Plato and Aristotle, proved to be exceptionally influential on

Western thought, so much so that it has been said that all Westerners are born as either Platonists or Aristotelians.

Plato and the Knowledge of the Cosmos

Plato was born in 427 B.C. in Athens, where he lived until after the turn of the fourth century. He then traveled to Italy, where he is reputed to have learned from the Pythagoreans there. Once back in Athens he established a school, which became known as the Academy. Here students could study and learn from the master himself.

The question of fundamental reality.   Plato agreed with predecessors such as Zeno when they argued that our sense experience of the world did not put us directly in touch with the foundation that undergirds the reality of the world. But when he examined the objects of experience, he was struck by two things. First, they possessed similarities that permitted them to be placed into classes. Second, while objects like cups were all similar to each other, no two cups were *exactly* alike.

Plato believed that it was impossible to choose one material cup and say that it represented the reality of the class of cups, since to do so would ignore its differences from other cups. On the other hand, if he said that only individual cups were real, then he was ignoring the similarities that permitted them to be grouped into clearly identifiable classes.

Plato's solution to this problem was to declare that what was real was the *form* of cups—something we might call *cupness*. This was the basic reality underlying all individual cups, a reality in which they all participated and that they expressed in their own ways. What was true of cups was true of every other object of sense experience. Each object was an imperfect representation of the fundamental reality of the form in which it participated. So if we want to know fundamental reality, we must attempt to get to the forms—and not be content merely with the objects—of sense experience. Experience presents us with diversity and change, but the realm of the forms is eternal and changeless.

Reason is the tool that allows us to acquire knowledge of the forms. It alone enables us to recognize the sameness that all cups possess. Plato acknowledged that our experience of individual objects could assist reason in its task. But we must understand that what observation provides is not true knowledge. Observational knowledge derives from the realm of changing things, not that of eternal forms. In the last analysis, reason must operate on its own.

Plato's cosmogony.   In a work entitled *Timaeus,* Plato set forth his cosmogony, that is, his account of the origin of the cosmos, the physical world known to the senses. He explained why even the world we experience was susceptible to rational explanation. He first made clear that the cosmos was not part of the eternal realm—it had not always existed, as did the eternal forms of reality. The cosmos had been created by an artificer, a divine workman.

Plato went on to say that when this divine workman created the world he looked to that which was eternal. At first the visible world was not at rest, but moved in an irregular and disorderly fashion. So the workman brought order out of disorder. Because the artificer was good, he framed the world "in the likeness of that which is apprehended by reason and mind and is unchangeable." The order undergirding the physical world, then, was unchangeable, even if the physical world itself underwent change in accordance with this order.

So in one important respect Plato disagreed with his predecessors. They had removed personal agency from explanations of nature's operations, arguing that nature's operations were simply the way things were, the inevitable and necessary natural course of things. For them the world's rationality derived from the underlying unity that tied all change together.

But for Plato the world could not be devoid of consciousness. It had been made in the likeness of *mind* and as a result it bore the qualities of mind. Mind, for example, was the source of intention and purpose; hence the cosmos bore the marks of purposefulness. In fact, for Plato the cosmos possessed a soul, which animated the world and caused it to move. Note that Plato did not wish to attribute every individual event in the world to the purposes of individual gods, as in the mythological accounts. In fact, the workman who had created the cosmos was not a ruler over it. He could not alter the arrangement or interfere with it. Like anyone else, he had to live in accordance with the plan he had imposed.

The structure of the cosmos.   Plato respected mathematical properties as expressions of divine reason, as had the Pythagoreans from whom he had learned. He therefore appealed to these properties when articulating the plan the divine workman had employed in creating the cosmos. He was aware, for example, that there were but five so-called regular solids, each of whose sides is an equilateral polygon. Plato associated four of these solids with the elements—earth, air, fire, and water— that had been identified by the philosopher Empedocles (ca. 492–ca. 432 B.C.) as the basic material constituents of the physical world. He associated the fifth regular solid with the overall cosmos.

Using this arrangement, Plato constructed a reasonable explanation for why and how elements combined to form the diverse substances we encounter in our experience of the world. For example, three of the five regular solids are made up of four, eight, and twenty equilateral triangles respectively. Plato could explain the formation of new substances by saying that triangles of the four elements were separated and recombined into new arrangements. Note that in this explanation substances are determined solely by the shapes that the combinations of triangles dictate. Thus, the nature we experience is determined by the properties of mathematics, and for Plato and his followers there was an expectation that descriptions of nature would utilize mathematics.

In Plato's description of the heavens the role of mathematics became very obvious. The outermost boundary of his cosmos—into which were fixed the myriad of stars of the heavens—was spherical. It turned around the central Earth, also a sphere, once every twenty-four hours, imparting to all heavenly bodies a twenty-four-hour, or diurnal, motion. In the *Timaeus,* Plato noted the existence of the seven stars that wandered against the background of the rest of the stars, each exhibiting its own so-called proper motion. Each wanderer, or "planetes," had a designated location, beginning with the moon and sun, which were closest to the Earth. He envisioned the planets as spherical and was aware that their proper motions occurred within the band of the zodiac, which was inclined at an angle to the axis of rotation of the celestial sphere.

Plato knew that each planet underwent its own kind of proper motion against the background of the stars. Some of these motions were quite complicated, involving changes of direction and speed. He also knew that these various motions could not be chaotic and unpredictable. If the heavens were rationally ordered, as he believed, then the planets had to be following a regular pattern of some sort. But which one? And how was it to be described?

It made sense to Plato and those who came after him that the accelerations some of the planets exhibited were apparent and not actual. Mars, for example, sometimes appeared to slow and reverse direction from night to night, only to soon resume its first direction. How to explain what was going on? To admit that Mars actually decelerated and accelerated seemed to Plato and the Greek astronomers to introduce a chaotic element into the heavens, something that ran completely against their philosophical assumptions. But if these motions were the resultant sum of individual uniform motions acting simultaneously, then the chaos of accelerated heavenly movements could be captured as a combination of these regular uniform motions.

According to historical tradition, Plato challenged the students of his academy to discover what combinations of regular uniform motions would result in the paths the planets were observed to follow. In coming up with solutions the students understood that they had to employ perfectly circular figures, the most rational of geometrical shapes. These two requirements—the use of circular paths for planets moving at uniform speeds—satisfied for them the conditions necessary to preserve nature's rationality. But could the right combination of circles be found that would result in the motions the planets were observed to undergo?

A student of Plato, Eudoxus of Cnidos (ca. 400–ca. 430 B.C.), offered the first solution to Plato's challenge. He assumed that the Earth was at the center of the cosmos and that the sphere of the stars lay at its outer extremity. This sphere, into which the stars were fixed, turned around the central Earth once every twenty-four hours and accounted for the observed daily motion of the heavens. To account for the motion of the planets Eudoxus created a set of interconnected spheres for each one, with the planet itself fixed onto the innermost sphere. For the sun and moon he used three spheres each, and for each of the remaining planets he employed four spheres. Including the sphere of the fixed stars, this amounted to a grand total of twenty-seven spheres.

Take for example the spheres used to explain the motion of the sun. The outermost of the three spheres rotated in a synchronized way with the sphere of the fixed stars and drove the other two in its wake. This outer sphere, then, explained why the sun shared the daily trip the stars took across the sky. The axes of rotation of the inner two spheres were positioned at the proper angles and rotated uniformly at the proper speeds so that, when combined with the motion of the first sphere, they permitted the sun to lag behind the stars along the zodiacal band as it was observed from Earth to do over the course of a year. Needless to say, Eudoxus was an excellent geometer to be able to combine the motions of the spheres in just the right way to obtain the desired resultant sum of the motions of the sun and the other planets as seen from Earth.

Aristotle's Worldview

One of the students who came to Athens to study with Plato was a young man of seventeen from Macedonia, a region that lay north of the Greek city-states. Aristotle (384–322 B.C.) became Plato's most famous pupil. He enjoyed an illustrious career, remaining with Plato for two decades before embarking on a period of travel and study after the master died. When his travels were over he came back to his native Macedonia, where he became tutor to the young Alexander, soon to become the great conqueror of foreign lands. He later established a school in Athens, where he spent most of his remaining days.

Aristotle's contrast to Plato.   It becomes clear that Aristotle had an independent mind when we examine how he differed from his teacher regarding the knowledge of nature. Plato influenced the development of Western natural science through his emphasis on the theoretical—the ideal order undergirding the natural world and the use of mathematics to describe it. Aristotle's contribution was different. He began with the individual existence of things and worked his way from the specific to the general.

Not surprisingly, Aristotle did not agree that Plato's eternal forms were somehow more real than the material world and existed apart from it. He refused to permit the separation of a sensible object's form from the substance of which it was made. For Aristotle, form and matter came together to constitute an object of sense. The iron in an iron ball supplied the matter, whereas the heaviness, spherical shape, hardness, and dark color were all attributes of form that gave the piece of iron its identity as a sensible object. Matter could not exist without form, nor could form exist without matter to serve as its subject.

Aristotle was no less convinced than his teacher that reality was ordered and could be rationally accounted for. But, except in certain areas that clearly lent themselves to mathematical representation (such as the mathematical description of the heavens), he did not want to begin at the level of the abstract. He preferred to begin with sense experience of actual things as they were and uncover from them whatever order there was to find. If Plato emphasized the theoretical approach to the investigation of nature, Aristotle represented its empirical side.

Aristotle agreed with Plato that the earliest Greek philosophers had erred when, in removing divine agency from their treatment of the cosmos, they ended up with a cosmos without purpose. But Aristotle's account of purpose in the cosmos was different from Plato's. Plato had traced the purpose that existed in the world to the mind of the divine workman. Aristotle located purpose in a thing's nature; that is, things change in a certain way because it is in their nature to do so. This explains why, for example, a baby grows or even why heavy objects fall. So, where natural objects are concerned, the cosmos is an inherently purposeful place, since all change comes about as the result of things acting according to their natures.

The world of living things.   Aristotle paid more attention to the individual objects of sense experience than did Plato, so his interest in nature ranged much more widely than had that of his teacher. This was evident in Aristotle's investigation of the world of living things. Unlike many in his day, Aristotle did not think of the study of animals, for example, as repulsive. He wrote *The History of Animals,* in which he described hundreds of kinds of animals and even tried to group many into classes.

Aristotle's treatment of organisms, which drew on his understanding of the matter and form of natural things, greatly influenced those who came after him. While the matter of the body of organisms was obvious enough, the form, which involved the defining properties of a natural object, was very special in the case of living things. Aristotle appealed to the presence of *soul,* which he associated with the form of living organisms. For example, it was in the nature of things with soul to undergo reproduction and growth, activities that inanimate objects did not display.

Aristotle noted three kinds of soul, each one corresponding to a different type of organism: vegetable, animal, or human. Plants reproduced and grew because of their nutritive soul. Animals added to these activities the capacity to react to stimuli because of their sensitive soul. Humans had a rational soul, which added the capacity for reason. These distinctions would last well into the eighteenth century among physiologists struggling with the unique challenges of understanding living organisms.

Heaven versus Earth.   Aristotle's treatment of the world of living things illustrates that he was an astute observer of the phenomena that he encountered on Earth. In the tradition of those who came before him, he also considered the heavens and what transpired there. But Aristotle insisted that the two realms, the terrestrial and the celestial, were qualitatively different places that should be kept separate. The dividing point between the two regions was the orbit of the moon. What occurred in the heavens had to be considered in isolation from the treatment of things below the moon.

The celestial realm for Aristotle was a place of perfection and eternity. What transpired there—for example, the movements of the stars and planets—followed eternally repeating patterns that never altered. As a result, nothing ever came to be or ceased to exist above the orb of the moon. To suggest that a new star could be born not only ran against what was observed, but it violated a principle that Aristotle had accepted from one of his early predecessors—the principle that something cannot come from nothing. This same principle was the basis for Aristotle's firm conviction that the cosmos itself could not have had a beginning. It must have always existed.

Because the heavens were completely different from the Earth, they were not made of the matter that was found on Earth. The matter constituting heavenly bodies did not consist of earth, air, fire, and water, but of a fifth kind of matter, a quintessence, found only in the celestial region. It did not make sense to Aristotle that there could be places where there was no matter at all, either in the heavens or on Earth.

Just as it was natural for the heavens to be as they were, so too was it with the region below the moon. Here, as we know well from experience, things do vary. This is the realm of change, the place of generation and corruption, where things come to be and go out of existence. Unlike the heavens, which exuded qualities of the divine, the terrestrial realm was anything but eternal and godlike.

When Aristotle examined the factors that gave rise to change he spoke of different kinds of causes that were necessary for change to occur. Each individual cause had to be present before change could take place, and knowing the individual causes was to understand why change occurred. From our discussion of both form and matter we can appreciate why he singled them out as two of the causes of change. For something to change it must take on new form, hence there was what Aristotle knew of as a *formal* cause of change. Since matter and form must exist together, the substance involved served as the *material* cause.

Aristotle noted two additional causes necessary for change. What he identified as the *efficient* cause was closest to what we consider a cause—the action that brought about the change. Finally, recall that Aristotle's cosmos was a very purposeful place. Natural purposes came from the nature of individual things. But agents, who supplied the actions that produced change, also acted with purpose. Whatever the source of the purpose, Aristotle regarded the end or goal of the change as its *final* cause.

We can illustrate the operation of the causes that produced change with the occurrence of a rain shower. The formal cause of the rain is the shape of the drops that water, the material cause, assumes. The efficient cause is the cooling of the warm air that has risen, which becomes water that falls as rain. The final cause here is the function rain serves to irrigate land so that things can grow. In the case of an artificially produced change, say the building of a birdhouse, the formal cause would be the shape of the house, the material cause the wood that takes on that shape, the efficient cause the person who constructs the house, and the final cause the builder's reason for making it—to give shelter to birds.

The constitution of natural objects.   Aristotle accepted the division of matter into four elements as set forth by Empedocles, but he did not regard these elements as the ultimate building blocks of natural objects. For that he chose qualities that we sense bodies to possess. Aristotle did not think of these qualities in the same way we might be tempted to. For example, it appears at first glance that he appealed to different degrees of temperature and moistness to explain the basis of the elements and different intensities of weight to determine where they naturally could be found. But when he spoke of hot versus cold, wet versus dry, and heavy versus light, he understood each member of a pair to be its own quality, the contrary quality of its opposite. It was not that something with the quality of coldness lacked heat. It had cold.

Each of the four elements resulted from a combination of the four qualities of cold, hot, wet, and dry. Cold combined with dry yielded earth; cold with wet gave water. Hot with wet produced air, and hot with dry resulted in fire. Nature, or for that matter humans, could alter existing combinations of the qualities to initiate transformations of one kind of matter into another. Take earth (cold and dry). If the dryness of earth could be displaced by wetness, earth could become water (cold and wet). This was what happened when a solid substance dissolved in a solution.

Aristotle, remember, did not think of hot and cold as degrees of one quality, temperature, but as individual qualities of their own. As such, however, these qualities *could* exist in various degrees. The great diversity of natural substances we observe in the terrestrial region resulted from various combinations of the four elements. The intensity of each individual quality making up an element also added to this diversity.

The qualities of heavy and light determined where a substance naturally was found. Consistent with his conception of the purposefulness of natural things, Aristotle believed that each natural object below the moon strove to be in its own place. This understanding is known as Aristotle's *doctrine of natural place*. It was the nature of objects with heaviness to strive toward the center of the cosmos, while those with lightness endeavored to recede from that center. Earth, the element with the greatest heaviness, displaced water, which possessed less heaviness. Likewise fire, with more lightness than air, displaced air as it receded from the center of the cosmos. Natural substances made of various combinations of elements assumed their location in the terrestrial realm according to the differing degrees of the fundamental qualities of heaviness or lightness they possessed.

How and why things move.   From the preceding discussion we can understand one of the reasons why natural objects might move. If they were displaced for any reason from their natural place and not forcibly retained, they would endeavor to return to that natural place. Carry a rock to the edge of a cliff and throw it off. With only air to restrain it, the rock would fall toward the center of the cosmos, stopping only when it was hindered by the presence of a substance with greater heaviness.

It was an axiom of Aristotle's physics that "all motion requires a mover." That meant that no motion ever occurred by itself, without a mover to cause it. In the case of the rock falling from the cliff, the mover was the nature of the substance, which acted in concert with the external air. This doctrine provided one of the reasons why Aristotle and his many disciples believed that the Earth did not move. What could possibly be its mover? Another reason arose from the doctrine of natural place. The substance that made up the Earth—the element earth—strove to be at the center of the cosmos, where it naturally assembled in a stationary sphere.

What about the motion that did not result from substances moving toward their natural places? What of forced motion or the motion living things exhibited? Where in these circumstances was the mover Aristotle required?

In the case of living things that could initiate locomotion, the soul was the mover. In violent motion the application of force was necessary. This could happen naturally, when two things collided, or artificially, when a force was deliberately applied to move an object. When we throw a spear at an advancing enemy, our arm supplies the force that moves the spear. The action of the mover in this case is obvious.

But what mover is acting once the spear leaves our hand? The spear continues to move, but it is no longer in contact with a mover. In this case Aristotle claimed that the spear set up a current as it parted the air. This current of air rushed in behind the spear to communicate continuing motion to the spear. Of course this accounted only for the horizontal motion of the spear. It would also eventually fall to the ground as the result of its attempt to return to its natural place.

If this explanation of projectile motion was less than satisfactory, Aristotle's further analysis of how things moved was quite impressive. He concluded that there was a relationship between the speed of the motion produced and the force that was applied. In modern terminology the speed and force were directly proportional to each other; that is, the greater the force, the faster the speed produced. He also noted that there was an inverse relationship between the speed and any resistance that had to be overcome. Thus, the greater the resistance, the slower the speed. This beginning of a quantitative analysis of motion was impressive enough to serve as the basis for subsequent commentary until the time of Galileo, in the seventeenth century.

The structure of the heavens.   In his descriptions of the terrestrial realm Aristotle did not want to force the details and complexity of what he observed into the abstract relationships called for by mathematical description. To do so would, in his mind, have missed much of the richness of nature's intricacy. Aristotle felt that geometry lost much of its certainty when it was applied to the phenomena of the terrestrial region.

Where the heavens were concerned, however, Aristotle did not hesitate to utilize mathematical description. This was a territory of perfection, where the purity of abstract relationships demanded nothing less than the precision of mathematical expression. He considered the system of spheres put together by Eudoxus to account for the movements of the stars and planets and made an important correction to it. In so doing Aristotle demonstrated his sensitivity to the demands of reason.

Eudoxus utilized twenty-seven uniformly rotating spheres to answer Plato's challenge to find a rational explanation for what had to be the merely apparent confusion of planetary motions. By embedding a planet in the innermost sphere of the set assigned to it, and by connecting that sphere to others whose motion was communicated to the inner sphere, Eudoxus was able to explain why the planet underwent the unique motions that it did. Were we on the planet itself, we would experience only the uniform circular motion of the sphere in which it was embedded. But, as seen from the Earth, the planet appeared to move in an irregular pattern that was produced by the interconnected spheres.

Aristotle no doubt admired the solutions Eudoxus had come up with for the individual planets. But he then realized that there was something fundamentally wrong with them. As seen from the Earth, Eudoxus's system accounted for the motion of the nearest planet, the moon, well. But the motion of the next planet, the sun, would be affected by the moon's motion if the spheres were nested as Eudoxus asserted. Eudoxus treated each planet in isolation from the others, as if the motion of one planet would not be affected by the spheres of all of the other planets between it and the Earth.

Aristotle inserted a correction into the system to take care of the problem. He inserted new spheres for the moon just beyond the original set Eudoxus had created. These spheres undid what the three spheres had accomplished, so when the line of sight arrived at the sphere carrying the sun the distortion caused by the moon's spheres had been removed. He did the same beyond the sun and the other planets as well. In the end his system made use of over fifty spheres.

By insisting on this correction Aristotle revealed that he was concerned with a higher degree of rational explanation than his predecessors. He demanded that the depiction of the heavens be internally consistent. It was not enough to solve the mathematical challenge the complex motions of the planets raised. Aristotle took seriously the implication of using spheres. Their reality as objects in the heavens could not just be ignored. This is another example of the perceptibility of this remarkable Greek philosopher.

◎ The Greek Heritage in Natural ◎ Philosophy After Aristotle

The accomplishments represented by Plato and Aristotle set the tone in philosophy for years to come. The issues these great thinkers articulated continued to be debated for centuries; indeed, they are still discussed today. But as great as their philosophical achievement was, it did not go unchallenged in the intellectual world of the late fourth century B.C. A major rival emerged in the person of Epicurus (341–270 B.C.), who further developed the theme known as *atomism,* which had been introduced in the late fifth century B.C. According to Epicurus, everything in the world, including the gods, was made of atoms. As they moved through space, the atoms collided with each other in ways that accounted for all that happened. This was a deterministic system in which no divine mind or natural purposefulness existed. Everything that transpired resulted from purely mechanical interaction. By basing their explanation and understanding on mechanical causation, the atomists created an approach that would hold enormous appeal for many natural philosophers who came after them, including many today.

Epicurean atomism was not the only view of the time that opposed the idea of a purposeful cosmos as described by both Plato and Aristotle. But all who argued in favor of determinism faced the task of rendering a convincing account of the apparent purpose encountered in experience—for example, in the human capacity to exercise free will. In spite of this substantial challenge, atomism persisted. It remained viable among some in the ancient world and its revival much later in the early modern era would continue to serve as an alternative to Platonism and Aristotelianism.

The Continuing Development of Mathematics and Astronomy

Astronomy, as we have noted, was cultivated in ancient times in isolation from the world of change beneath the moon. Because it dealt with a realm of perfection, the heavens, it could be described by the abstract properties of mathematics. For this reason we should regard astronomy in this period and all the way to the early modern era as a branch of pure mathematics. Solving the problems of astronomy for Greek thinkers was a mathematical problem. In spite of Aristotle's concern with the disruption caused by interacting spheres, the physical constitution of the heavens played a less important role than their mathematical structure prior to the late sixteenth century.

There were some areas where mathematics was applied in the terrestrial realm, including optics and the law of the lever. The Greek mathematician Euclid (ca. 300 B.C.), author of the famous geometry book, *Elements,* also wrote a book entitled *Optics,* in which he used geometry to develop a theory of visual perspective. The most famous mathematician of the time, however, was Archimedes (ca. 287–212 B.C.)

Archimedes and the law of the lever.   Archimedes of Syracuse is best known for the legend that he exposed chicanery in the making of the king's crown by showing that the amount of water displaced by the crown was not the same as that displaced by an equal weight of gold. However improbable this event may be, Archimedes clearly was a man of genius. He is credited with numerous inventions and a wealth of mathematical insight. In his book *On the Equilibrium of Planes,* we can directly observe his application of mathematics to the lever.

Archimedes argued that geometrical symmetry can serve as the foundation for proving that when the weight times its distance from a fulcrum is equal on both sides of a fulcrum, a beam will be in equilibrium. He began by placing three equal weights on a beam, one directly over a fulcrum located at the middle of the beam, and the other two at the ends. Symmetry dictates that the beam will remain level. He then moved the weight that was above the fulcrum halfway to one end and compensated for this move by bringing the weight from that end to the same position. He now had a combined weight that was twice as heavy as the single weight at the other end, but located half as far from the fulcrum. Archimedes reasoned that these compensating moves left that side of the beam with the same capacity to balance the other side that it had before.

From this result Archimedes was able to generalize to the conclusion that equilibrium is achieved when force times distance to the fulcrum is equal on both sides of the fulcrum. Thus, a small force times a large distance can achieve the same effect as a large force over a small distance. It is possible, therefore, to move a very heavy object by placing a fulcrum close to it, inserting a lever over the fulcrum, and pushing down on the other end of the lever with a much smaller force. According to later testimony, Archimedes was reputed to have said: "Give me a place to stand and I will move the world."

Ptolemy's astronomy.   The greatest astronomer in the period after Aristotle was Claudius Ptolemy, whose contributions came in the second century A.D., some five hundred years after Aristotle's death. Ptolemy came from Alexandria in Northern Africa, although he had no relation to the Ptolemies who once ruled in Egypt.

When Ptolemy compiled his system of astronomy in a work later called the *Almagest,* he consulted and described the astronomical work done by those who had preceded him. He has thus become an important source of information about the work of some figures from between his time and that of Aristotle. Through him we have become acquainted with the new directions Greek astronomy was taking in the period before the birth of Christ. Ptolemy also added his own contribution to the solution of the main problem confronting the astronomer—how to devise a mathematical description of the movement of the heavens that best fit what careful observers saw happening there.

Ptolemy wrote, for example, about Apollonius of Perga, a Greek mathematician from the third century B.C. According to Ptolemy, Apollonius devised new mathematical techniques to describe planetary motion that improved the fit between what was described and what was observed. Even though observations were only as accurate as the naked eye permitted, astronomers became more and more demanding. They became less satisfied with the fit obtained using the techniques of interlocking homocentric spheres employed by Eudoxus and Aristotle. The new techniques Ptolemy described afforded much greater flexibility to the astronomer to adjust the mathematical depiction of the motion of the planets. One of these new techniques was the use of an eccentric point, a point positioned off-center.

The eccentric point permitted a ready depiction of the apparent slowing down and speeding up of a planet as it underwent its proper motion from night to night against the background of the stars. Consider the figure here, which shows the Earth (E) located at a vastly exaggerated eccentric position to the center (C) of the planet's orbit. As the planet travels in uniform motion around the circle, it will *appear* when viewed from Earth to go faster through the distance from A to B than through the longer distance from D to F. This is because an observer will note that the planet sweeps out angle AEB quicker than it does angle DEF (since the distance from A to B is less than that from D to F), even though the angles are the same size.

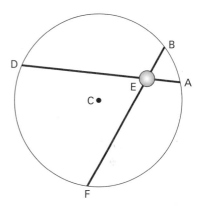

Eccentric Point

The introduction of another device, the epicycle (a circle on a circle), made greater sense of the apparent backward and forward movement over time of planets such as Mars. Here the assumption was that the planet (P) traveled uniformly around a smaller circle whose center (C) moved around the central Earth (E). The epicycle was a very flexible tool. It could be used to make minor adjustments in a planet's orbit, or to generate extremely complicated paths. Assume, for example, that in this figure the planet P moves uniformly around C faster than C moves around the Earth (E). Viewed from Earth, the speeds of P around C and C around E will add together when the planet is at P, giving the planet the appearance of an acceleration.

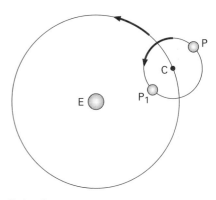

Epicycle

When the planet moves to the position P₁, however, its more rapid speed around C will dominate over the slower motion of C around E, so from Earth the planet will appear to be moving backward. Eudoxus had also been able to account for this retrograde motion of a planet, but the epicycle was a simpler explanation.

The epicycle was more useful than merely explaining the retrograde motion of planets. Depending on how big the epicycle was in comparison to the orbit of C around E, and depending on the relative speed assigned, extremely complicated paths could result. This provided the ancients with a tool of great flexibility when trying to simulate complex motions that were observed in the heavens.

From Ptolemy's viewpoint these devices had improved the fit, but they had not yet made it good enough. He wanted to do still better, so he introduced what would turn out to be a highly controversial new tool called the equant point. He employed this point in conjunction with the eccentric; that is, he located a second off-center point on the opposite side of the center from the first eccentric point, where the Earth was positioned. He then referred uniform motion around the circle to this second point, not to the center of the circle. This meant that as the planet traveled around the circle it passed through equal angles in equal amounts of time. However, the constant angular velocity was not referred to C, the center of the circle, but to the equant point. In this figure, for example, the planet passes through angle AQB in the same amount of time it takes to pass through angle AQF. But clearly it has farther to go from F to A than it does from A to B.

If it took the planet the same amount of time to go both distances, it must have been going faster from F to A than it was from A to B. This meant that its motion was no longer uniform throughout its path. By using an equant point, especially in combination with an eccentric point and epicycles, Ptolemy was able to improve the fit between the mathematical description of planetary motion and how the planets were observed to move.

More than others, Ptolemy did concern himself with how the physical constitution of the heavens might impact the theoretical explanations he had devised. He asked, for example, how the crystalline spheres in the heavens would have to be arranged in order to produce something like the epicycle. In this particular case he imagined a small sphere, in which the

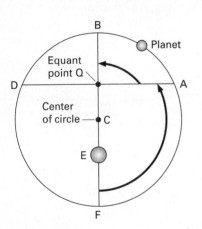

Equant Point

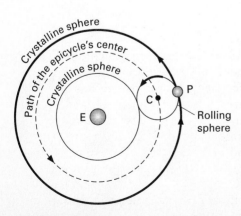

Crystalline Spheres

The Problem with Ptolemy's Equant

Ptolemy's equant presented those who came after him with a dilemma that sometimes faces natural scientists today. By using the equant point he was able to depict and predict planetary motions more accurately. But is a new approach to a problem justified wholly by its usefulness? Is the natural philosopher free to choose *any* explanation as long as it enables him or her to give a more accurate rendering of observational data?

The problem for those who came after Ptolemy was that the equant implied that the planet actually accelerated and decelerated, and that ran directly against the Greek assumption that only uniform motion could exist in the heavens. Ptolemy, of course, could argue that he had retained uniform motion, that he had merely reinterpreted how it was to be understood. But, in the view of many who came later, permitting a planet to experience accelerated motion ruined the rationality of the explanation. To them, reinterpreting uniform motion to refer to the equant point was cheating.

In the sixteenth century one of the major motivations for Nicolaus Copernicus's new heliocentric system was to rid astronomy of the equant. He believed that the power driving the motion of the planets had to be constant. The prospect that it was fickle was one before which, he said, "the mind shudders." Actual acceleration might work better, but it eliminated the possibility of giving a rational explanation in the mind of Copernicus and those of his day. As we shall see in Chapter 5, Copernicus's greatest achievement was, in the eyes of the scholars of his time, to have removed the equant point from astronomy.

There are no easy answers to the questions posed here. The problem generated by Ptolemy's device has haunted natural philosophy throughout its history. On encountering a twentieth-century version of the challenge, a famous physicist once remarked that to grasp what nature is like we might have to change what we mean by *understanding*. To do that, however, is bound to generate opposition among those committed to a tried and true method. The investigation of nature, then, can present us with challenges that can lead to basic questions about what it is that we are actually doing.

planet was embedded, rolling around a channel between two larger crystalline spheres and carrying the planet with it as it went. As the outer large sphere rotated, its motion forced the small sphere to turn, generating the planet's proper motion and the epicyclic effect.

In spite of this attention to the physical spheres in the celestial realm, however, Ptolemy did not permit physical considerations to dictate or eliminate the mathematical descriptions. In the end he was content to have come up with a mathematical explanation that accurately depicted the motion of the planets. And this attitude persisted for centuries after him. While many variations of his system were created, consisting of different combinations of eccentrics, epicycles, and equant points, Ptolemy's basic approach lasted until the time of Galileo.

A Burgeoning Tradition in Medicine

The earliest Greek medicine shared a heritage similar to that of early Greek natural philosophy. Just as the first philosophers distinguished their explanations of nature from mythological understanding by removing divine agency from the world's operations, there were those who began to resist invoking the gods as the cause of individual disease. In both cases, however, the new tendency to appeal to natural process existed alongside a continuing tradition of mythology.

Medicine had its Pythagoras in the person of Hippocrates of Cos (ca. 450–370 B.C.), although he lived about a century later than the famous mathematician. Like Pythagoras, Hippocrates is a shadowy figure about whom we know very little. But—again like Pythagoras—his work gave rise to a group of like-minded followers and his name is widely known even today because of a development attributed to him, the famous Hippocratic oath. According to the original oath, medicine was to be practiced by those initiated into the holy mysteries, another similarity to the secrecy of the Pythagoreans.

For all that, there is a major difference between the Hippocratic writers and the Pythagoreans. While the latter incorporated the divine into their view of reality, the followers of Hippocrates presumed that bodily processes, health, and disease could be described as natural phenomena, without reference to arbitrary, supernatural involvement.

The authors of the Hippocratic writings commonly felt that health resulted from the proper balance among key bodily fluids, sometimes called humors. Initially three humors were identified: blood, phlegm, and bile. Disease resulted when the balance among the humors was disturbed as, for example, in winter, when the body excreted excess phlegm when sick with a cold. Excess bile was associated with summer vomiting, and too much blood was expelled in instances like nosebleeds or menstruation. The physician should imitate nature and remove excess blood to treat associated illnesses. Later physicians included a fourth humor—black bile (in contrast to bile, sometimes identified as yellow bile). Imbalance here produced melancholy.

The preferred treatment among the Hippocratic physicians, however, was to prescribe the right diet. Diet was not restricted to what patients put into their mouths, although proper food and drink was essential. It involved their entire lifestyle—how much sleep they had, how much exercise they did, how much fresh air they enjoyed. Keeping a proper diet, then, could prevent or cure illness by balancing what the body required.

Physicians attempted to learn about the body's internal parts by dissecting animals. Following the death of Alexander the Great, in the Hellenistic period, physicians in Alexandria were said by later writers to have shifted from dissecting animals to humans. This development revealed the internal structure of the body in a way that was unknown earlier. Herophilus of Chalcedon (ca. 330–260 B.C.) investigated male and female reproductive anatomy, distinguished between nerves, veins, and arteries, and described the basic structure of the brain. Others contributed as well to this new exploration of human anatomy.

At the time of Galen of Pergamum (ca. A.D.129–200), who was born in Asia Minor, human dissection was not permitted, as it had been earlier in Egypt. Forced to work from animal dissection, Galen nevertheless became the primary ancient authority whose influence dominated Western knowledge of medicine and anatomy for many centuries.

Galen is perhaps most well known for his anatomical and physiological work, even though he acknowledged that he had to infer the internal structure of human bodies from their counterparts in animals. He divided physiological functions into three parts. First, there was the brain, which governed the nerves. As the blood passed through the brain it received spirit, which the nerves carried to all parts of the body, enabling sensation and muscular stimulation. The other two parts had to do with the flow of blood. Galen concluded that there were two systems of blood flow. One began in the left side of the heart, carrying blood enlivened by the heat of the heart through the arteries to the rest of the body. The source of the other was the liver. From here nutrients were carried through the veins to other parts of the body, including the lungs, by way of the right side of the heart. Between the heart's right and left chambers was the septum wall, through whose invisible pores Galen claimed blood seeped.

From Plato's *Timaeus* Galen took the notion of a divine workman who made the cosmos according to the dictates of a rational order. Just as the heavens reflected this eternal order, so too, argued Galen, did the body of living organisms. He found the natural purposefulness of the world Plato endorsed in the adaptability of the animal and human body to the functions it performed. Although Galen himself was not a Christian, his celebration of the evidence of a wise creator that was to be found in the anatomy of living things made his work understandably appealing to Christian natural philosophers for a long time. This attraction provided another reason why Galen's authority grew to enormous stature in the West.

In the Roman Empire after the beginning of the Christian era Greek learning continued to set the tone. Rather than embark on their own intellectual pursuit of theoretical knowledge, the Romans took their cue from what Greeks had and were still accomplishing. As time passed and knowledge of the Greek language waned, some Roman scholars undertook translations of Greek works so that they would remain accessible. When Romans themselves delved into natural philosophy they did so as a leisure activity, not, as the Greeks had, out of an appreciation of the value of intellectual achievement for its own sake. As a result, the natural philosophy that marked the early Middle Ages made little pretence to being as original as what great Greek thinkers had done.

Early medieval natural philosophy has been characterized as popular and encyclopedic, meaning that it aspired to producing compendia of information that all educated people found fascinating. From handbooks of natural knowledge to compilations of information about natural history, mathematics, and geography, the Romans wished to keep their feet on the ground and not to become lost in abstruse thought that neither provoked their curiosity nor served some obviously useful purpose.

Where early Christian thought was concerned, the attitude of Augustine (A.D. 354–430), a bishop of the church who lived in North Africa, was mildly sympathetic to the merits of natural philosophy. While he emphasized the need to focus on spiritual matters, he did not want Christians to be dismissed because they were ignorant of knowledge about the material world. As a result he recommended that Christians study nature to the extent that natural knowledge promoted a hearing for the message of salvation. Augustine's view assigned natural philosophy the status of a handmaiden to religion—it remained in the service of something more important than itself.

The heritage of the Greek accomplishment moved from the Latin West to the Islamic world, where it underwent further development before it was retrieved again by Western scholars. But to a remarkable degree, what the ancients had achieved defined for the future the questions to be asked about nature, the problems to be solved, and the kinds of answers that would be permitted.

Suggestions for Reading

James Evans, *The History and Practice of Ancient Astronomy* (Oxford: Oxford University Press, 1998).

Alexander Jones, ed., *Ancient Science,* Volume I of *The Cambridge History of Science,* ed. David C. Lindberg and Ronald L. Numbers (New York: Cambridge University Press, 2007).

David C. Lindberg, *The Beginnings of Western Science* (Chicago: University of Chicago Press, 1992).

G. E. R. Lloyd, *Aristotle: The Growth and Structure of His Thought* (Cambridge: Cambridge University Press, 1968).

CHAPTER 2

─────────────◎─────────────

Learning in the Middle Ages

At the beginning of the fourth century of the Christian era, a Germanic people known as Visigoths, from the delta of the Danube River, began to invade northern Italy. By A.D. 410 they were successful in their attempt to take Rome. And this was not the only time Rome was sacked in the fifth century—it occurred again at mid-century. The lack of political and social stability that resulted from these wars meant that the pursuit of intellectual matters, and even the cultivation of education, took a back seat to more pressing concerns.

At the same time, the growth and development of Christianity as a religion brought with it a new institution—monasticism. Monasteries were places where those who wished to devote themselves to Christian holiness could focus on that task. To the extent that monasteries were self-supporting, they functioned as separate communities. This meant that in addition to requiring practical knowledge about such mundane matters as carpentry, food production, or the growing of herbs for medicines and spices, monks were responsible for whatever formal education was provided in the monastery.

For the most part, education was directed toward spiritual edification. There were exceptions, but in general the heritage of Greek natural philosophy did not hold great appeal in monasteries, where the primary goal was spiritual better-ment. Monks consulted classical thought and sought new knowledge about nature only when it served a higher purpose, namely, the understanding of spiritual truth. But that is not to say that monasteries made no contribution at all to the intellectual development of the Latin West. Although monasteries differed about which secular works might be relevant to spiritual development, those with the broadest vision made sure that Greek works were translated into Latin. Through this activity, through the study of scripture, and eventually through the development of theology, the monasteries kept literacy and scholarship alive.

It must be said, however, that in the period from A.D. 400 to 1000 there was no intellectual activity on a par with that of earlier Greek natural philosophy. We look in vain during this time for a Roman Ptolemy or Archimedes, figures distinguished by their grasp of earlier scholarly work and by their creative genius in developing new ideas about the natural world. Further, many Greek works on natural philosophy and mathematics were not translated and thus dropped from view.

◎ Transmission of Greek ◎ Learning to the Near East

The early development of Western Christianity drew its organization from the basic structure of the Roman Empire; that is, while local churches were self-governing, there emerged by the fourth century a structure that was modeled on the organization of the Empire itself. In this structure groups of churches were arranged into provinces, with preference given to churches of the large cities. By the fifth century there were five major patriarchies, the two most important being Rome, the oldest center of the Roman Empire, and Constantinople, to which the Emperor Constantine moved the seat of the Empire in 330.

In spite of their common religious origins, the development of the Eastern and Western churches reflected differences in language and outlook. For example, Latin became the universal language of liturgy in the West, while in the East the church

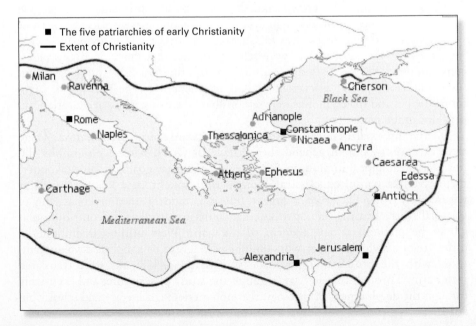

The Five Patriarchies of Early Christianity

tended not to require Greek, the major language of the Eastern church, but to defer to the vernacular language of its various ethnic regions. As a result, a so-called Orthodox Catholic tradition emerged in the East, under its patriarch, to counterbalance the emerging papal authority of Roman Catholicism in the West.

The Nestorians

The great diversity represented by the eastern patriarchies was bound to lead to internal dispute about Christian belief. In particular, disagreement arose about the nature of Jesus Christ himself. Did he have one nature that was wholly divine and wholly human at the same time, as some insisted, or did he have two separate natures, one human and one divine, that existed alongside each other? Nestorius (d. ca. 450), who was the Greek-speaking patriarch of Constantinople, supported the latter interpretation. When this view was formally condemned by a church council in 431, Nestorius retired to a monastery in Antioch and was subsequently banished to Egypt.

There were many who agreed with Nestorius's understanding of Christ's nature, however. When another church council condemned his teaching in the middle of the fifth century, his followers fled the Roman Empire and gathered first in Syria and then in Persia, where they were able to flourish. The Persian king was greatly interested in Greek culture and philosophy and over time the Nestorians, who supported the king's interest, rose to positions of importance.

Nestorian activity in Persia was responsible for the transmission of a great deal of Greek learning to the Near East. As was common in the Eastern church, the Nestorians deferred to the indigenous language in their liturgical practice; in this case it was Syriac, the language of Persian culture. As they continued to introduce Greek ideas and customs into the region, they translated Greek works into Syriac. This was the initial stage of the spread of Greek natural philosophy to the Near East.

Islam and Natural Philosophy

With the birth of Muhammad in Mecca in the late sixth century the destiny of Persia and many other lands of the Near East was substantially altered. Muhammad aggressively spread his new religion of Islam throughout the Arabian Peninsula and began to move to regions beyond. In Persia the new Muslim leaders granted protection to the Nestorians and they in turn continued to exert a significant cultural influence under their new rulers as the guardians of Greek learning. In the middle of the eighth century one of these rulers, al-Munsur (754–775), built a new capital in the city of Baghdad, which became known as a center of intellectual activity.

Beginning with the reign of al-Munsur and extending through his successors, Baghdad became a translation center where Syriac and Greek works were systematically brought into the Arabic language. Nestorian Christians were not the only people engaged in this massive translation effort. Muslims, Jews, and other Christians, known as Sabians, were also involved in what Islamic rulers regarded as an important task. By the turn of the first millennium of the Christian era there was very little of

Plato, Aristotle, Galen, and other authors of the Greek corpus that had not been made available in Arabic. Greek thought was cultivated to a sufficient degree to preserve a tradition that in many cases had been neglected in the West.

The attitude of Arab leaders toward Greek learning was in an important respect similar to that of the early Romans; that is, they were interested in foreign ideas primarily to the extent that they were useful. Mathematics and astronomy had obvious benefits, as did the knowledge gleaned from Greek medicine. The same could be said, perhaps to a lesser degree, about astrology and even natural history. Greek philosophical treatises proved relevant, and therefore useful, to Arabic theologians. As a result of this utilitarian outlook, the status of Greek learning always remained secondary to the heart of Islamic culture and religion.

The magnitude of the effort of translating so many Greek and Syriac sources into Arabic might suggest that translation defined the extent of the Islamic achievement. But Islamic thinkers did not simply take over Greek ideas without comment or question. While they acknowledged their debt to those who had come before them, Islamic authors sought to improve and correct the material they had inherited.

Islamic astronomers made lasting and innovative improvements to Ptolemy's system. They made new star charts, they introduced trigonometry into astronomical calculation, they corrected deficiencies in the Julian calendar, and they perfected the design of the astrolabe, an instrument used to measure the position of stars in the sky. In mathematics Islamic scholars introduced a system of numerals that they had inherited from India and then perfected. It included a zero as a placeholder, removing the sometimes misleading and ambiguous designations of numbers in existing numeration systems. They also made use of symbols for unknown quantities in an exceptional work from the ninth century with the Arabic words *al-Jabr* (from which we have *algebra*) in its title.

In the study of optics an Islamic scholar named Ibn al-Haytham, also known as Alhazen (ca. 965–ca. 1040), produced a new theory that was critical of the existing assumption that vision involved light streaming out from the eye. Alhazen argued that light entered the eye from the outside. He then brought together mathematical, physical, and medical ideas for the first time to explain how vision works.

Another area of knowledge that Arabic scholars developed was alchemy. Early Christians in Hellenistic Egypt pursued *chemia,* the practice of refining metals. Islamic thinkers were interested in the Greek texts on this subject, which they translated under the rubric *al-chemia.* They wanted to learn whatever they could about how material substances interacted because that could be useful in the production and purification of substances and in the coloring of alloys to look like gold and silver. As they pursued the subject on their own, Islamic thinkers improved practical techniques for distilling, crystallizing, and sublimating natural substances.

More than two thousand alchemical works are attributed to one Jabir ibn Hayyan, whose actual existence is in some doubt. Many of these writings originated from the long period between the middle of the ninth century to the end of the tenth. They contain the viewpoint that metals originated from combinations of two exhalations found underground. Because all metals shared a common origin, it was theoretically possible that one could be transmuted into another. To carry out this transmutation, the early Greeks attempted to prepare a substance called *xere,* which the Islamic authors translated as *al-iksir* and which later was known as the philosopher's stone. This very old process, then, became a focal point for the alchemical tradition that would later come to the West from Islam.

From the high point of translation activity, roughly between 900 and 1250, natural philosophy continued to flourish and develop in the Islamic world. By the thirteenth century internal strife and Mongol attacks from without led to instability and a decline of interest in natural philosophy. By the end of the fifteenth century, as those in the West were realizing the existence of unknown lands, little remained of the Islamic fascination with nature. However, the period of assimilation of Greek thought into Islam that began with al-Munsur at the end of the eighth century and started to decline in the thirteenth marked a span of over five hundred years during which the Islamic world remained the world's center of natural philosophy.

◎ Revitalization of Intellectual ◎ Pursuits in the West

In approximately 1000, priests and monks, or "those who pray," constituted one of three groups that made up the emerging political, economic, and social system known as feudalism. The other two were the knights who defended society, or "those who fought," and the almost 90 percent of the population known as "those who work." European society was agrarian—towns and cities were just beginning to emerge in the Northwest of the continent. For the most part farming was done communally in connection with a manor.

By the twelfth century change was evident. As they consolidated noble status for themselves knights began to define it in terms of lineage. For over two centuries powerful individuals had been emerging as kings, with lesser knights assembled around them at court. New agricultural techniques permitted the expansion of arable land and the growing of new crops. New villages and towns had sprung up, making possible the development of urban centers marked by trade and commercial activity. With these new urban centers came a new social group, the bourgeoisie, who provided the leadership for urban political institutions.

If these developments had been underway for some time, so had changes in attitudes toward intellectual pursuits and toward the role of education.

The Enlivening of Learning

In the early monasteries the cultivation of intellectual ideas was centered on religious and theological matters, so natural philosophy was virtually ignored. Occasionally a translation of a Greek work on philosophy appeared, but what interest it held stemmed from its relevance to theological study. With the emergence of a powerful leader in Europe in the latter half of the eighth century, however, education became a greater priority.

Flourishing of the schools.   Charlemagne (742–814) consolidated under his rule several areas of what is now continental Europe. By the time of his grandsons, virtually all of Europe, with the exception of contemporary Spain, had become part of the Carolingian Empire. Charlemagne concluded that as the leader of the state he himself would benefit from a broad education and he believed that educational reforms would strengthen the church as well. To head the school he wished to create he brought Alcuin of York (ca. 730–804) to his court at Aachen. Alcuin had been the head of the cathedral school in York in the British Isles. Together with his new teacher, Charlemagne proceeded to reform education throughout the realm, which, with the exception of Spain, included the greater part of the European continent.

It would be a long time before the literate monks outnumbered the illiterate monks. But through the establishment of new cathedral and monastery schools, education became a much more important enterprise than it had been before. As a result of broadening the content of education, some works on natural philosophy and mathematics began to reappear.

These changes were in part merely the flourishing of the venture begun under Charlemagne. Schools in the new urban centers naturally expanded the curriculum to match the new interests emerging around them. Even cathedral schools, which retained their religious focus, increased the subject matter investigated whenever it could be made relevant to their mission. A decisively greater willingness to rely on human reasoning marked the investigation of both traditional and new subject matter, especially in the urban schools. Of course scholars who sought to gain insight into religious truth by rationally examining the basis of Christian doctrine ran the risk of opposition from those who did not trust this method of philosophy.

The attraction of Platonic thought.   Another aspect of the enlivening of learning in the eleventh and twelfth centuries was a renewed appreciation of the classics. Charlemagne's grandson, Charles the Bald (823–877), supported the translation of several neo-Platonic Greek works into Latin. This attraction to Platonic thought marked the initial phase of translation activity in the West, and would increase significantly within a few hundred years. In the twelfth century Plato's *Timaeus*—which had been translated in the first half of the tenth century—and other Latin works whose neo-Platonic perspective could be reconciled with a Christian outlook, provided a framework in which ideas about the creation of the physical world through the work of a divine craftsman could be understood.

An important feature of this blending of Plato with Genesis was the introduction of what the historian of medieval natural philosophy David Lindberg has called "the new naturalism" of the twelfth century. By this Lindberg refers to the restriction of divine activity to the initial moment of creation. Scholars more and more assumed that from the time of creation on, things in nature, including human beings, followed a regular course dictated by natural causation. Humans represented a microcosm of what occurred in the cosmos writ large—the macrocosm. Both were tied together by the common structure they shared.

In the *Timaeus* the divine craftsman was bound by the order he created—he could not intervene to change it at will. In twelfth-century naturalism that claim was weaker. It was not that God *could not* intervene into nature's order, merely that God customarily did not do so. Nevertheless, by establishing the importance of nature's regular course, twelfth-century scholars opened themselves to new questions about how that course worked.

Translation of Works into Latin from Arabic and Greek

In the closing years of the eleventh century Europeans from all ranks of society joined together in response to a call from Pope Urban II to attack the Muslims who had besieged the Eastern church and who controlled Jerusalem. This and subsequent crusades provided contact between West and East, even if it did nothing to promote an appreciation of the achievements of Islamic thought and culture. But an awareness of the accomplishments of Islamic natural philosophy was not entirely absent. Men such as Gerbert (ca. 945–1003), who had traveled to northern Spain a century earlier to study Arabic mathematics and astronomy, enriched the Western perspective.

Christian forces began to be successful in driving Muslims from Spain during the eleventh century. Although it was not the only supply of information about Arabic natural philosophy and mathematics, Spain was, in fact, an important source. Not only was it close to other European countries, but there were also Christians in Spain who lived under Muslim rule. They proved to be of enormous value as Europeans sought to absorb Islamic thought.

Many Arabic works in the areas of medicine, natural philosophy, mathematics, alchemy, and astrology came into the Latin language in this fashion. Since Islamic scholars had taken Greek works as their original point of departure, classics from Ptolemy, Euclid, Aristotle, and Galen came into Latin through Arabic. At the same time, Latin scholars became aware of original Arabic works on subjects such as medicine and algebra.

Yet not all of the Greek classics came into Latin by way of this flood of translations from the Arabic that occurred in the twelfth century. Some were made directly from Greek texts that had been preserved in southern Italy. Once begun, the demand for translations continued to grow. By the end of the thirteenth century a great deal of the Greek and Arabic works on natural philosophy and mathematics had been recovered in the Latin West.

The Appearance of Universities

The existing educational establishments were not immune from the climate of social and intellectual change that was occurring in the eleventh and twelfth centuries. More and more young people wished to take advantage of the expansion of resources coming into the Latin West. In response to what was happening, a new institution—the university— made its appearance during this time. It differed from the urban school, which typically had one master and a dozen or two students, by bringing multiple masters together in one place.

The oldest university was founded in Bologna, Italy, around the middle of the twelfth century, followed by the University of Paris (1200), Oxford University (1220), and ten others over the course of the thirteenth century. By 1420 fifteen more had come into being, all but three of them during the fourteenth century. The explosion of the student population during this era called for a comparable expansion of the teaching force. Around the turn of the thirteenth century, for example, there were over seventy masters teaching at Oxford.

In the medieval university there were three higher areas of learning, or faculties, to deal with the advanced subjects of theology, medicine, and law. Before students could prepare themselves in one of these faculties, however, they had to study the arts. Traditionally, Roman education had consisted of seven liberal arts, which were divided into two groups known as the trivium and the quadrivium. Grammar, rhetoric, and logic formed the trivium, while arithmetic, geometry, astronomy, and music made up the quadrivium. The medieval university retained this basic division of the arts, although it placed more emphasis on logic than was common earlier.

The more technical subjects of the quadrivium were no more emphasized in the university than they had been in the urban schools. But it would be misleading to infer that concern with the natural world therefore received very little attention. On the contrary, the physical and metaphysical works of Aristotle, plus commentaries on these works, had come into Latin beginning in the second half of the twelfth century and in the thirteenth, exactly the time when the universities had come onto the scene. The great breadth of Aristotle's thought captured the attention of many scholars, and Aristotelian natural philosophy became a central feature, even a compulsory subject, of the university curriculum.

The early universities varied in size, from large centers such as Paris, which had over two thousand students, to small schools with a few hundred. There was considerable turnover among the students; many left after a short time for a variety of reasons. Students arrived at the age of thirteen or fourteen with a grammar school knowledge of Latin. Those who remained would take an examination for a bachelor's degree at age seventeen or eighteen from the master under whom they had studied. If they passed they could study for three more years to earn a master's degree, which gave them membership in the arts faculty with the right to teach. New masters were sometimes required to teach the arts for a certain period before they could opt to attend one of the higher faculties (theology, medicine, law) for another five or six years.

The medieval university was an important contributor to the development of Western scholarship in general and natural philosophy in particular. Medieval university scholars preserved and passed on to the West the Greek and Arabic wisdom that had accumulated for over a thousand years. They shared a dedication to systematically applying Aristotelian logic to the body of knowledge they had gained from both sacred and secular sources. They encountered new challenges when their investigations opened up new territory that had to be understood in light of received truth. In general they responded to these challenges, not by dismissing secular ideas

Medieval University Professors

whenever they ran counter to Christian revelation, but by creatively attempting to develop new perspectives that reconciled discrepancies. They undertook, in other words, a reasoned articulation of human meaning and in so doing they put in place an approach that would shape the Western mind for centuries to come.

◎ The Assimilation of Ancient ◎ Knowledge of Nature

Of the many different kinds of texts that were translated into Latin, those on more technical subjects such as mathematics and astronomy stood out as far superior to anything that was known previously. As a result, such works ran little risk of generating controversy. The same could not be said for Aristotle's thoughts about the cosmos because they involved claims that were different from received wisdom. And it was Aristotle's works that had become favorites of the scholars in the new universities.

Aristotelian Natural Philosophy and Christian Belief

Some of Aristotle's conclusions ran contrary to widely held religious beliefs. When this occurred it was necessary either to oppose them or to figure out a way to reconcile them with existing doctrine.

Problems raised by Aristotle's philosophy. In Chapter 1 we learned that a fundamental principle of Aristotelian philosophy was that something could not come from nothing. For Aristotle that meant that the cosmos could not have had a beginning. Matter and the elements must have always existed. He also believed that matter would never go out of existence—the cosmos would not have an end.

This is just one example of an Aristotelian tenet that did not blend with traditional Christian thought. The Bible clearly indicated that God had created the world, although it was less clear whether the authors of the scriptures intended to convey the idea that God had created an ordered world out of preexisting chaos or that he had created it from nothing. By the medieval period, however, the principle of *creatio ex nihilo* (creation from nothing) was a commonly understood doctrine. Such a reading of the book of Genesis ran directly against Aristotle's belief that something cannot come from nothing.

DEVS DE NICHILO CREANS VNIVERSA

NI CHIL

God Creating the Universe

There were other points at which Aristotle's natural philosophy clashed with Christian doctrine. His understanding of nature's regularity highlighted the Greek tendency, evident in the earliest Greek philosophers, to remove or downplay the role of personal agency in explanations of natural processes. Not all Greek thinkers displayed this tendency. Plato's divine craftsman, for example, could be reconciled with the idea of a purposeful creator, even if the God of medieval Christian theologians could intervene in nature's workings and Plato's craftsman could not.

But, as we saw in Chapter 1, Aristotle agreed with Plato's criticism of the early Greek philosophers that, in removing personal agency from nature, they had also mistakenly removed purpose from the natural world. Like Plato, Aristotle insisted that there was purpose in nature, but he did not account for it by appealing to the role of a divine craftsman or any other external personal agent. Purpose for Aristotle came from the nature of things themselves. Effects followed causes in accordance with the natural purposes that reigned in nature. Aristotle appealed to the activity of the divine—in his notion of the unmoved mover—only to begin the chain of cause and effect.

As a result, Aristotle could sound very deterministic to Christian thinkers of the thirteenth century. The most blatant contradiction to Christian understanding was the denial of the possibility of miracle that was clearly implied in Aristotle's thought.

Aristotle's impersonal god was not responsible for the purpose in nature and certainly could not interfere with the cause and effect relationships it had set into motion. When believers prayed to God, they counted on his having the power to act on their requests. If Aristotle was right, their God was impotent, not omnipotent.

Aristotle's teaching about form and matter also supplied a problem for the university scholars who had turned to the translations of and commentaries on his works. According to Aristotle, form and matter could not exist apart from each other; there could be no form without matter and no matter without form. The form of the living body entailed more than its shape—it involved the organization necessary for life to exist. So for Aristotle the soul, which provided this organization, was the material body's form. Because form and matter existed together, when the body no longer continued to exist, neither did the soul. How could this viewpoint be reconciled to a belief in the immortality of the soul?

Reconciling Aristotle and Christian doctrine.   Aristotle's thought presented scholars of the thirteenth century with a real dilemma. In spite of these problems, they were very much attracted to the rich diversity of content of the Aristotelian corpus. Beginning with his treatment of logic, Aristotle enticed late-medieval thinkers to make use of their reasoning powers to undertake investigations of their own. Further, Aristotle had considered so many different subjects that the agenda he had set would keep scholars busy for generations to come. Nothing comparable to the wealth of information and analysis was available anywhere else. It was difficult to remain unimpressed.

There were therefore numerous attempts throughout the thirteenth century to explain how, in light of the obvious problems it created, Aristotle's thought was to be understood. One of the first university masters to lecture on Aristotelian natural philosophy was Roger Bacon (ca. 1220–1292) at the University of Paris. Bacon's approach was to emphasize the utility of the new philosophical texts, including Aristotle's, that were coming into the Latin West. In addition to the obvious ways in which, say, astronomical knowledge was helpful in making the church calendar accurate, Bacon declared that the new philosophical knowledge allowed a better understanding of scripture and was useful in confirming Christian doctrine.

Some were not as confident that potential problems could be ignored. They too enjoyed applying their rational abilities to understand God's word and God's world, especially with the new tools that had become available from pagan sources. But they showed an awareness that adopting the starting points of Aristotle—for example, that something could not come from nothing—might take their rational analysis to conclusions that challenged accepted belief.

Where Aristotle's system was concerned, they identified the eternity of the world, determinism, and the mortality of the soul as unacceptable conclusions that flowed from false starting points. They were not confident, as Bacon appeared to be, that the use of pagan philosophy was automatically beneficial. They trusted starting points that in their view had been divinely revealed to them, not those adopted by pagan Greeks.

The question therefore arose in the medieval university as to whether philosophy could be trusted. When philosophy was taken to mean Greek philosophy, it

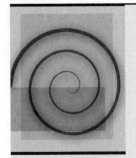

Delimiting Faith and Reason

The NATURE of SCIENCE

There was already precedent in the late medieval period for designating the different roles played by reason and faith. In a slogan for which he is well known, the early-twelfth-century archbishop, Anselm of Canterbury (1033–1109), declared, "I believe in order that I may understand." Anselm was aware that understanding, which depends on the use of reason to tease out the implications of an assumption, cannot be attained unless there is some assumption at the start. Reason only goes to work once that assumption is in place.

The role of reason here can be likened to the operation of a computer once the return key has been struck. The computer carries out an analysis based on software that dictates the contours within which the reasoning takes place. To carry the analogy a bit further, in natural philosophy the computer's hardware, including its read-only memory (ROM), might be likened to the physical universe with its "hard-wired" laws.

If there is merit in this analogy, it would suggest that the frequent positioning of faith and reason as opposites is misleading, since they must work together. Sometimes, though, the assumptions on which rational analysis is based are so commonly shared that they become invisible. In such cases—for example, when assumptions are widely shared in the scientific community—it may incorrectly appear that scientific analysis is devoid of any assumptions and that reasoning in science therefore stands in opposition to the embrace of assumption.

encountered substantial suspicion and criticism. But Thomas Aquinas (ca. 1224–1274), the doctor of theology at the University of Paris, argued that the problem did not lie with philosophy per se. He conceded that philosophy as a means of investigating God's truth could not justify its starting points on its own—it could not *by itself* make known the truths God had given by revelation. But, given those starting points, philosophy would never oppose them.

Why bother with philosophy at all, then? Aquinas argued that while philosophy would never contradict the truths of revelation (since it assumed them at the outset), it might nevertheless lead to new truths that had not yet been realized. Further, philosophy provided an excellent means of showing what was erroneous about critiques of Christian doctrine.

Thomas Aquinas constructed an elaborate synthesis of Aristotelian and Christian ideas, one that eventually became ensconced in medieval thought. To do this he had, of course, to explain where Aristotle was in error and where he was just misunderstood. The Thomistic synthesis, as it has been called, involved a great deal of subtle analysis, some of which became so involved that many readers did not follow it. In many cases Aquinas used philosophical speculation to uncover options most people had never realized were possible. For example, he disagreed with Aristotle's position that the world was eternal. But he suggested that eternal matter would be just as dependent on God for its existence as was matter that had come into existence at some point. Thus, Aristotle's position was wrong—but not absurd.

The Reception of the Thomistic Synthesis

Aquinas appears to have felt that it was a mistake to oppose reason to faith, philosophy to religion. Philosophy had mainly to do with reasoning from a set of starting points, not, as in religion, with the establishment of those starting points. The two should not be opposed because they functioned for different purposes. They could only be set in opposition if they were working to accomplish the same end.

Others, like Siger of Brabant (ca.1240–1284), took a narrower view of philosophy, focusing on the starting points themselves. Siger concluded that the two *should* be kept rigorously opposed to each other *because* their starting points were different. He was impressed not so much that philosophy and religion were doing different things, but that religion adopted assumptions of one kind, philosophy of a different kind. In religion the defining starting point involved a God above nature; in philosophy the starting point was nature itself.

Given this variation in the perception of the relationship between religion and philosophy, faith and reason, it is no wonder that church authorities grew suspicious of the role of philosophy. This was especially true because those who argued that philosophy was completely separate from religion felt free to engage in philosophical reasoning regardless of the conclusions that resulted. When their philosophical analysis produced outlandish conclusions that clashed with accepted doctrine, they simply declared their loyalty to church doctrine. Meanwhile the radical inferences of their philosophical reasoning circulated in their writings.

Church authorities felt that they could not stand idly by without taking a stand. Encouraged by Aquinas, the bishop of Paris condemned thirteen philosophical conclusions propagated by those who claimed that, because philosophy was separate from religion and subservient to it, they were free to draw whatever philosophical conclusions they wished. Aquinas died in 1274, but three years later 219 additional philosophical propositions were condemned, including many that Thomas himself had defended. Needless to say, his reputation suffered a blow.

Many of the condemned propositions had to do with what were regarded as limitations on God's power. For example, Aquinas had made it clear that he agreed with Aristotle that life as we know it did not exist elsewhere. But the bishop of Paris understood him to mean that there *could not* be life elsewhere—for the kind of philosophical reasons that Aristotle had given. Aquinas had in fact been careful to say that God could have created life elsewhere, but that he had not done so. Nevertheless, he was misunderstood. "That the First Cause cannot make other worlds" was one of the condemned propositions. With this condemnation the church put itself on the side of the possible existence of extraterrestrial life, reversing the judgment of earlier Christian theologians that such a pagan belief was unacceptable. It also opened up new questions for theologians and natural philosophers to ponder—whether, for instance, Christ's sacrifice covered beings that might live on other worlds or whether he would have to go to those worlds to die again. The consideration of such matters lasted well into the nineteenth century and forms a fascinating episode in the history of the interaction between astronomy and religion.

Aristotelian natural philosophy did not fare well in the condemnation of 1277. The bishop viewed Aristotle's assertion of the impossibility of some natural arrangements—his denial, for example, that a vacuum could exist—as a limitation of God's omnipotence and he condemned Aristotle's claim accordingly. But the ban on teaching any of the condemned propositions, on pain of excommunication, was less significant for what it prohibited than it was as an indication of the state of the intellectual world of the late thirteenth century. While it was clear that the recovery of ancient philosophical works had made their way into the minds of many university scholars, it was just as clear that not everyone agreed with this assimilation of ancient wisdom. In fact, the leaders of the church expressed the greatest opposition.

Aquinas's reputation and his blending of Aristotelian thought with Christianity soon recovered from the disrepute brought in 1277 by the bishop of Paris. But the relationship between philosophy and religion would remain a contentious issue for some time to come. As we will see in the next chapter, by the fourteenth and fifteenth centuries new objections were raised about the ability of philosophy to assist in the articulation of Christian truth.

◎ Late Medieval Natural Philosophy ◎

The natural world envisioned by medieval thinkers bore many resemblances to that of the ancient Greeks. By the same token, there were valuable new contributions made by university scholars that corrected and otherwise altered what had been received from the past. In many cases the new ideas about nature or mathematics emerged in the context of theological discussion, always a subject of primary concern at this time in Western history. There are important ways, then, in which medieval theology helped to shape the content and, even more significantly, the assumptions that would support the Western heritage in natural philosophy and, eventually, in natural science.

Change, Forms, and the Study of Motion

It seems curious to our modern minds that the early Greek philosophers spent so much effort discussing the question of whether change was possible. But the concept of change is profoundly fundamental to natural philosophy. Without some kind of change nothing would happen, and the universe could not be perceived to exist. How and why things change therefore became basic questions of Western natural philosophy. To answer these questions it was often less important to know what is changing than to focus on the *process* of change. For example, a medieval consideration of the problem of change might consider how grace or charity could be increased in a person. From the examination of such questions some remarkable conclusions about change emerged.

Medieval scholars followed Aristotle in acknowledging different kinds of change. In the heavens the only change was of place, as the stars and planets moved uniformly with their various spheres. Below the orb of the moon, in addition to change of place, things could exhibit changes in size or in the degree of the qualities they possessed. Finally, change occurred in the terrestrial realm when a thing came into or went out of existence.

The intension and remission of forms.   Medieval scholars were interested in an intriguing question about changes in form, specifically, when a change intensified or weakened. When a form or quality intensified—for example, a person's charity—did the increase occur in definite increments or did it do so continuously? One twelfth-century theologian suggested that charity was a gift of God; hence, when it increased, it did so in discrete amounts as the result of a specific act, not as a process. But in the next century another theologian asserted that charity could be *gradually* strengthened over time.

The theologians asked themselves how the form "flowed" from one state to the next. In the first case, when charity increased suddenly by a definite amount, the flow of change in the form occurred because the quality was first present in one degree and then in another. Here the flow is like that observed in the floor indicator lights of an elevator as they move from 1 to 2. The flow in this case is due to the light being first in one position and then in a completely different one. Even though it seems to, the light does not move from one location to the other; rather, it goes out at one point and reappears at the other. One state is replaced by a second state. When we see the light move from 1 to 2 we create the flow in our minds from our observation of one light going out and the other coming on. So a person whose charity increased in this way first had one amount, then, as the result of God's gift, a measurably greater amount. Any perception of flow in the increase of the charity was created by the observer.

In the second case, when the flow of change in the form is continuous, the flow is like the ascent of a person in a swing from a lower position to a higher one. Here the flow can be said to be real—it is inherent in the person who experiences the increase in height above the ground. If charity increased this way there would be no detectable difference from one moment to the next, but over time an augmentation would become evident.

With this admittedly esoteric discussion, medieval thinkers were treating a problem that would persist in natural philosophy—namely, the problem of continuous versus discontinuous change. The general subject behind the notion of flow in the intensification and remission of forms is the question of motion, which soon became an object of investigation. In the fourteenth century, for example, a remarkable group of scholars at Merton College of Oxford University came up with new concepts to characterize what was happening when things moved.

Because the Oxford professors were curious about the notion of the intensity of a form—the degree to which the form is present in a body—they asked what the intensity of *motion* might be. The ancients had talked about motion solely in terms of the distance traveled during a specific time. As a result of insisting that motion must have intensity, medieval scholars created a new parameter—velocity. This new quality was an abstraction to be sure, one of many such medieval concoctions. From it soon followed notions of uniform motion (constant velocity), nonuniform motion (acceleration), and even uniformly nonuniform motion (constant acceleration).

The Oxford scholars went beyond merely defining these new parameters. They developed propositions that employed the new concepts in their explorations of motion. They knew, for instance, that a body that was uniformly accelerated would cover a certain distance in a certain time. They established a proposition: The distance covered when a body is uniformly accelerated over a specified length of time is the same the body would cover if it moved at its average velocity for the same amount of time.

On the basis of the achievements of the Oxford scholars, others soon took an important new step—the mathematical representation of the new parameters. A professor named Nicole Oresme (ca. 1320–1382) at the University of Paris made major contributions in this area. Following the ancients, who had used lines of differing lengths to represent varying amounts of space and different durations of time, Oresme began by representing degrees of velocity by different lengths of a line. A line half as long as another represented a velocity half as great as the other.

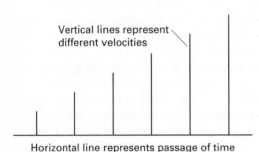

Vertical lines represent different velocities

Horizontal line represents passage of time

Use of Horizontal and Vertical Lines to Show Acceleration

His next step was to realize that he could also create a geometrical representation of acceleration. To do this he followed Aristotle in representing time by a horizontal line. He then placed vertical lines, which represented the increased velocities, at places along the horizontal line. In this way he was able to represent what the velocity was at the various points in time. He thus created a picture of what was happening in accelerated motion. If he drew a line connecting the tops of the vertical lines (not shown in the figure), it could be thought of as a representation of the acceleration.

Oresme could now use his new technique to prove the Oxford proposition concerning the distance covered by a uniformly accelerated object. Constant velocities were depicted by equal vertical lines along a horizontal time line, as indicated in the left-hand figure below. From the different velocities of the accelerated motion he chose the velocity at the halfway point, or the average velocity. Oresme conceived of the areas of the figures generated as a measure of the distance traveled. Since the two triangles emphasized in the right-hand figure below are equal, the area of the rectangle formed by the constant velocities is equal to the area of the triangle made by the accelerated motion. In other words, the distance traversed by a body moving for a

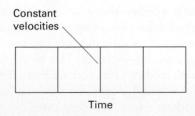

Constant velocities

Time

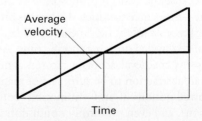

Average velocity

Time

certain time at uniformly accelerated motion is the same as that traversed by a body moving for the same time at the average velocity.

The cause of motion.   Apart from the descriptions of the various kinds of motion that exist, there is another basic question about motion that demanded the attention of ancients and medievals alike. It had to do with what causes motion. Aristotle had said that motion was caused by a mover; in fact, it was an axiom of his treatment of motion that all motion required a mover. Medieval scholars accepted Aristotle's axiom and they carefully considered his explanations of what the mover was in various cases of motion. They also responded to extensive Islamic commentaries on what the Greek master had set down.

A perceived problem almost since the time of Aristotle was his identification of the mover in cases of the unnatural motion of a projectile. In Aristotle's view such motion required the presence of an external mover because the projectile possessed neither a soul nor was its motion initiated by acting to seek its natural place. He had said that whatever supplied the power to launch the projectile was its mover, as long as that mover was in contact with the projectile. Once released, the projectile was moved by the medium through which it traveled. This was because the original propelling power moved not only the projectile, but also the air around it, imparting to the air the capacity to act as the mover from the point of release on.

Objection to Aristotle's requirement of an external mover for projectile motion had been raised as early as the sixth century. It was proposed then that the so-called violent motion of a projectile resulted from a motive power that was impressed on the projectile by the launching power and was afterwards *internal* to the projectile. This revision of Aristotle's teaching on motion came down through Islam to the Latin West, where it was initially rejected, later to be developed by the Parisian professor Jean Buridan (ca. 1300–ca. 1360).

Buridan suggested that when an arrow is shot from a bow, the bowstring impressed a power on the arrow. He called the impressed power *impetus,* which, he said, continued to act as long as it was not corrupted by resistance. In the case of projectiles the impetus moved the body, but as the motion was resisted, for example, by whatever medium it was moving through, the impetus was diminished until finally the projectile stopped.

Buridan's modification of Aristotle's explanation of projectile motion is but one example of the aggressive program of medieval natural philosophy. The professors of the medieval university did not simply accept whatever Aristotle had said; they actively sought to correct the master when they encountered what they regarded as insufficient or erroneous explanations.

The Heavens and the Earth

Medieval thinkers took care to respect any specific reference in the Bible to the Earth and its place in the cosmos. That did not prevent them, however, from considering possibilities not explicitly covered in Holy Writ.

The structure of the cosmos.   For the most part, a simplified view of Aristotle's understanding of the structure of the heavens became the dominant view of the medieval thinkers. That is, the cosmos was centered on the Earth and was surrounded by eight concentric spheres that carried the seven planets and the fixed stars. Clearly such a depiction of the heavens paid no regard to the proper motions the planets displayed, even though Aristotle himself had taken pains to account for planetary movement. He had, as we learned in Chapter 1, been aware of the combinations of spherical motions used by his predecessors and had inserted extra spheres to provide an internal consistency that the earlier systems had lacked.

If, in the common understanding of the structure of the heavens, medieval scholars ignored complicated motions, they also created new questions about that structure that were based on their Christian outlook. For example, the Genesis account of creation distinguished between the origin of the heavens, which occurred on the first day, and the firmament, which came into being on the second. In addition there was a reference to waters that existed above the firmament. To account for the various regions indicated in these passages, some medieval commentators inserted two extra spheres beyond the firmament or eighth sphere of the fixed stars—the first to provide a place for the waters above the firmament, and the second to designate a habitat for God and the angels.

Possible motion of the Earth.   The state of astronomy during the late medieval period did not rival the achievement of Ptolemy or the Arabic astronomers who came after him, but that is not to suggest that there were none who knew of Ptolemy's work or that there were no creative ideas at all. Both Jean Buridan and Nicole Oresme at the University of Paris revived the ancient question of whether a rotating Earth might produce the motions we observe the heavens undergoing. Buridan argued that if the Earth rotated on an axis it would not affect astronomical calculations, since astronomers dealt only with relative motions anyway. He concluded that it was impossible to decide the question on astronomical grounds alone, although it could be determined on physical grounds. Shoot an arrow into the sky. The arrow is not left behind by the Earth rotating under it, so the Earth must not be turning.

Buridan's younger colleague Oresme did not accept the argument about the arrow. He reasoned that if we look at the various motions that take place when sailors move on a ship, they all occur at the same time that the ship is moving through the water. He drew a parallel between this movement of the ship and that of the atmosphere above the Earth into which the arrow was shot. The motion of the arrow up and down could occur at the same time that the atmosphere turned with the Earth around an axis. Although Oresme appreciated the advantages of permitting the Earth to rotate, there were scriptural passages that seemed to rule it out. He explained away some of these passages as accommodations on the part of the writers of scripture, but in the end he deferred to the Psalmist, who declared that God had established the world, "which shall not be moved" (Psalm 92:1).

Heavenly influences on Earth and its inhabitants.   Everyone knew that the sun exercised a physical influence on the Earth. From this obvious knowledge it was but a small step to the more general assumption that the heavens and the Earth were physically linked to each other. The endorsement of a physical influence from the stars by ancient authorities like Ptolemy came down through Islam to the Latin West and remained unquestioned by medieval scholars and by their successors in the early modern era. To deny astrological influence would have seemed as ridiculous as denying that the seasons were linked to the annual motion of the sun or that the tides had something to do with the motion of the moon.

Astrological influence on human behavior was another matter. It was not that there were no defenders of such a link between the macrocosm and the microcosm. From the objections that were raised to this kind of influence it is clear that many believed in it. The opposition, present since the early Middle Ages, insisted that nothing impaired human free will. If astrology was seen as determining human affairs, it could not be accepted. This attitude remained well ensconced throughout the late medieval period and into the fifteenth and sixteenth centuries.

A modified position emerged during the twelfth and thirteenth centuries with the acquisition of ancient and Islamic works that discussed astrological influence. Although the stars did not affect the human exercise of free will, they could influence certain natural aspects of the lives of humans. For example, comparable to the kind of physical influence that seemed obvious where the seasons were concerned, there were likely arrangements in the heavens that affected a person's health and temperament. If the influence disrupted the health of whole communities, plague could result. It is therefore easy to understand why the study of astrology became a recognized part of the practice of medieval medicine.

Alchemy

The subject of alchemy came to the Latin West from Islam. The medieval theologian Robert of Ketton, who brought the Koran into Latin, translated an Arabic work on alchemy under the title *De compositione alchemiae* (*On Alchemical Composition*) in 1134. Although this was the first such work, many would follow. Within 250 years an enormous number of alchemical materials were accessible. In spite of this substantial growth of interest, alchemy was never part of the curriculum in the medieval university. This was because there developed, especially in the fourteenth century, a reaction against the subject among many theologians and philosophers, forcing those who pursued it to do so mainly on the margins of medieval intellectual society.

Late medieval opposition to alchemy was also dependent on Arabic thought, although medieval scholars believed that these criticisms had come from Aristotle. Around 1200 a section from a book by the early-eleventh-century Persian philosopher Avicenna (980–1037) was translated and inserted into an existing translation of one of Aristotle's works. Avicenna's critique of the transmutation of metals was thereafter routinely assumed to be by Aristotle himself, lending great credibility to its worth.

Avicenna's objection to claims that metals could be transmuted revolved around the difference between a natural and an artificial substance. In his view art was regarded as inferior to nature; hence those who attempted to transmute an inferior metal into a nobler one through an artificial process were doomed to failure. Further, according to Avicenna, the defining characteristics that made a metal unique were not accessible to the senses; rather, they existed deep in the nature of the metal. An alchemist could not get at them since he did not know what they were.

In response to this kind of criticism, which was frequently adopted in the universities, there were those who favored alchemy and disagreed with the limitations placed on art and on the human ability to transform the world. They also rejected the claim that humans could not know the defining characteristics of metals. In 1266 Roger Bacon argued that alchemy investigated matters of which Aristotle was ignorant, such as the way the elements gave rise to precious stones, minerals, and humors. Therefore, alchemy was a more fundamental subject than natural philosophy because it dealt with the elements as generators of new substances, whereas philosophy merely combined the four elements.

In the late thirteenth century the most significant alchemical work of the Middle Ages appeared, the *Summa perfectionis (Sum of Perfection)*, most likely written by an Italian named Paul of Taranto. He took the name Geber in a clear attempt to identify with the well-known alchemical figure in Arabic lore, Jabir ibn Hayyan. Geber provided extensive information about minerals and metals, how to work with them, and how to transmute one into another. To account for the properties of things Geber appealed to the smallest parts of their substance, which, incidentally, he did not think of as indivisible atoms. He explained, for example, that particles of a substance that were smaller than those of another could be packed tighter; hence such substances resisted the action of acids better than those that were packed together more loosely.

Alchemical practice was always vulnerable to exploitation by those with less than perfect integrity. Fake demonstrations of the transmutation of lead into gold convinced more than one investor to trust swindlers. With the development of legal restrictions placed on alchemical practice, the study of alchemy moved to the fringe of society during the course of the fourteenth century.

But interest in the subject did not wane; in fact, the official hostility to alchemy did not prevent its spread into the practice of medicine. Besides the transmutation of metals, another major goal of alchemists was to find healing medicines, even a substance that could prolong life. These two hopes, transmuting base metals into nobler ones and finding an elixir of life, would remain lasting and powerful goals of the Western alchemical tradition until well into the eighteenth century.

Medicine

Many of the general attitudes regarding medicine and its practice from the early years of Christianity were still in place in the late medieval period. The Christian West by and large had resolved early on the opposition between religious and secular understandings of medicine. For most Christians there was no contradiction between the perception of sickness as divine punishment or a vehicle of spiritual growth, and the view that illness was due to natural causes and should therefore be treated by natural means. God could, after all, work through natural means and the discoveries of effective treatments could be seen as divine gifts. While some still decried reliance on medical knowledge, most accepted it as useful.

Ancient works on medicine, however, suffered the same general fate in the early Middle Ages as other classical works of learning; that is, only a limited number of translations into Latin were available. To the extent that cultivation of medicine as a practice occurred at all, it took place largely in the monasteries, where care of the sick was seen as an expression of religious concern.

Because the writings of the Hippocratic and Galenic traditions were part of the vast corpus of Greek works that were translated into Arabic beginning in the eighth century, they and other Greek medical sources were preserved. Islamic medicine built on this Greek inheritance of medical wisdom by assimilating it into an existing indigenous tradition. Since almost all of the Greek medical texts were translated into Arabic, the materials that came down to the late Middle Ages from Islam represented a rich and diverse storehouse of medical knowledge.

The flourishing of urban schools provided the first context in which a Western medical tradition really began to take shape. Urban society produced elites who demanded people skilled in medicine to balance the proliferation of practitioners of all kinds that graced the landscape. From the wise woman who knew how to cure warts to the lithotomist who could "cut for the stone," the choices of a healer to treat a wide variety of medical ailments were legion. As the urban school gave way to the university, a process of differentiation developed that enabled recognized social distinctions to be made among healers.

Unlike alchemy, the study of medicine did find an institutional home in the university curriculum. This was an important development in Western medicine because it became an academic discipline, shaped by the Aristotelian natural philosophy to which it was exposed. Eventually an arts degree, which required the mastery of Aristotelian natural philosophy, became a prerequisite to medical study and medicine assumed a place among the higher faculties with theology and law.

But not all medical practice received the benefit of institutionalization among the intellectual elites. More than most aspects of intellectual culture, medical knowledge could not be confined to one social class. Virtually everyone in society was interested in the causes and remedies of disease, where relatively few expressed any interest at all in, for example, knowledge about the motions of the heavens. Because of this widespread interest in sickness and its treatment, there was not a radical difference between the medical ideas that circulated in society at large and those that were taught in medieval universities.

The central conceptions, widely shared, were that health depended on balance and that disease resulted from imbalance. The Greek understanding of the body's humors, or fluids, was easily grasped by anyone; hence, there was a basic conviction that a disturbance of the balance among the body's humors was responsible for sickness. Clearly, then, restoring balance was the fundamental treatment to be undertaken. Most often the assumption was made that the balance had been disturbed by the presence of an excess of one of the humors. To restore balance the extra fluid had to be removed; hence bloodletting and purgatives were the most common forms of treatment. Medievals also appreciated the information about the use of plants and herbs that came down to them from ancient authorities such as Dioscorides (fl. 50–70), especially those that induced vomiting or an evacuation of the bowel.

A variety of healers also stood ready to answer the call when a medical situation called for surgery. Depending on the procedure undertaken, the patient could be treated by an executioner or barber and even in some cases by a university-trained practitioner. In general, however, physicians with university degrees did not practice surgery because it was most often viewed as a craft that required technical expertise, not the mastery of medical theory.

We will consider the state of anatomical knowledge in the medieval university in Chapter 4, when we compare it to the important new developments that took place in the early modern period. Suffice it to say here that the anatomical work of Galen that first became known in translation from the Arabic during the late medieval period was not particularly helpful for instructing students. Only later, when more complete writings of Galen became available, did the study of anatomy take on greater importance.

The Bequest of the Middle Ages

More than any other single thing, the recovery of the writings of Aristotle shaped the content of late medieval natural philosophy. This fact should not overshadow the criticisms, revisions, and extensions of Aristotle's thought that took place first in the Islamic world and then in the Latin West. What eventually resulted was, of course, a Christianized version of Aristotelian natural philosophy. But at its foundation the categories in which nature was conceptualized—form, matter, and formal, efficient, material, and final cause—remained thoroughly Aristotelian. Because purposeful activity was exhibited throughout the natural world, the metaphor for nature was organism.

As they refined the description and treatment of the natural world, always integrating the results into their deeply religious outlook, medieval scholars set an agenda that lasted well beyond them. Many of the questions they asked, and to a surprising extent the kind of answers they permitted, persisted into the era that succeeded them.

Just as the growth of urban centers profoundly altered regional society in the Middle Ages, so too would the expansion of the known world that followed the late medieval period require new ways of envisioning nature and its inhabitants. Once again the writings of the ancients, including non-Aristotelian works, would play a part. The watchword of the coming era would be innovation.

Suggestions for Reading

Edward Grant, *The Foundations of Modern Science in the Middle Ages* (Cambridge: Cambridge University Press, 1996).

Edward Grant, *God and Reason in the Middle Ages* (Cambridge: Cambridge University Press, 2001).

David C. Lindberg, *The Beginnings of Western Science* (Chicago: University of Chicago Press, 1992).

David C. Lindberg, ed., *Science in the Middle Ages* (Chicago: University of Chicago Press, 1980).

CHAPTER 3

―――――――◎―――――――

Early Modern Innovations

Dynamic changes began to emerge in the fourteenth and fifteenth centuries in the Latin West. On the intellectual level the recovery of ancient texts captured the attention of scholars, making them rethink the synthesis of Christianity and Aristotelian thought Thomas Aquinas had crafted. The invention of printing, which increased the dispersion of new ideas, contributed to the growing awareness of newness. And with explorers penetrating farther and farther into unknown waters, knowledge of the Earth itself expanded in unanticipated ways, opening up whole new worlds to the European imagination.

By this time the landscape of power was changing as well. The most obvious difference from earlier times was the erosion of the power of the papacy. In earlier years the pope, as the leader of a unified Western Christianity, exercised considerable political power. But with the growth of the power of kings, whose interests were regional and secular, respect for papal authority weakened.

The disruption of the Vatican's power became openly transparent during the 1300s and 1400s. Philip IV of France actually captured Pope Boniface VIII at the beginning of the fourteenth century in the course of a dispute over the king's right to imprison a bishop. As the result of Philip's action, the papacy moved from Rome to Avignon in France in 1309. From then until 1377 the popes, all French, were located in Avignon during what is known as the Babylonian Captivity. No sooner had the papacy been relocated to Rome near the end of the century when another crisis arose. In hostile reaction to decisions of the new pope, Urban VI, cardinals fled Rome and declared the papal election invalid. They then proceeded to elect a second pope, who simultaneously ruled from Avignon and produced a schism in church leadership that lasted until 1417. At one point during this so-called Great Schism there were three popes, none of whom recognized the legitimacy of the others!

◎ Recasting the Medieval Intellectual Heritage ◎

The recovery of manuscripts from the past did not result in a permanent resolution of the differences between pagan and Christian notions about the cosmos or about nature's importance. Rather, it produced a continuing reexamination of existing ideas, a growing intrigue with the past, and even a reformation of faith.

Erosion of the Thomistic Synthesis

In a chaotic situation such as the Great Schism the appearance of new ideas was hardly surprising. One area of thought that felt the effect of intellectual change was the synthesis between faith and reason that Thomas Aquinas had forged in the thirteenth century. Especially in the fourteenth and fifteenth centuries, Aquinas's confidence that rational philosophical analysis was able to clarify and confirm the contents of revealed theological truth came under question. As we shall see, some of these new attitudes about the relationship between philosophy and theology had an impact on the manner in which mathematics and natural philosophy were viewed.

William of Ockham's new direction in philosophy. Among those questioning the relationship between faith and reason was William of Ockham (1285–1349). Born in England, he received a degree from Oxford before earning his master's degree in Paris. A Franciscan, he chose a vow of strict poverty based on the claim that Christ and his apostles had owned no possessions. This brought him into dispute with Pope John XXII and, as a result, he spent some years in prison.

Ockham was skeptical of the ability of natural reason to prove Christian doctrines such as the immortality of the soul. It was not that he wanted to completely separate theology and philosophy; rather, he argued that philosophy could not demonstrate the conclusions of faith with certainty. Therefore, the tenets of revealed truth had to be accepted through faith alone.

These conclusions were consistent with Ockham's view of philosophy in general. He believed that, especially where nature was concerned, philosophy had to do with propositions, not with things. Philosophy dealt with what humans know and that is a product of the mind. He believed that it was a mistake to regard philosophy's objective as an examination of things that exist. If that were the case, then philosophy would never be able to deal with generalizations, or what he called universals. It would have to remain on the level of individual things. Philosophy did have to do with universals, he taught, but they exist in the mind and only in the mind.

The study of motion provided an example of the implications of Ockham's view for natural philosophy. For him the noun *motion* did not refer to a real thing. The only real thing was the moving body, which existed in successive places. When we create an idea of the body's existence in these successive places and call it motion, we have created a name that does not refer to a real thing. It is just a name. Ockham's view is therefore called nominalism. His wish to cut away what was not needed and to retain only the fewest assumptions necessary is known as Ockham's "razor."

Ockham represents an important new direction in natural philosophy. In his skeptical view of the traditional claims of philosophy he moved away from what is known as realism, which claims to deal with whether or not something exists, to epistemology, which inquires about what is known in the mind. The significance of this distinction is very important for understanding what is meant by natural scientific truth in later centuries.

Nicholas of Cusa and learned ignorance. An example of the continuing skeptical tendency in theology from the fifteenth century is Nicholas of Cusa (1401–1464), a German theologian who eventually became a cardinal in the Roman Catholic Church and a legate for two different popes. In the late 1430s he composed his most well-known work, *On Learned Ignorance,* whose ostensible purpose was to clarify the relationship between God and humankind. In the course of making his case, Nicholas drew on the reader's knowledge of mathematics to illustrate his conclusions.

According to Nicholas, in acquiring new knowledge we compare what is unknown to what is certain. So in mathematics, for example, when we can trace a conclusion back to its origin in something we know to be true, we hope to establish the truth of our conclusion as well. This is obviously impossible, however, in all considerations of the infinite. The infinite has no comparison to anything we know, because human knowledge deals only in the realm of the finite. So clearly humans could have no knowledge about God, who is infinite in comparison to us.

But even when we compare finite things we run into difficulty. Nicholas agreed with Pythagoras that things are constituted and understood through number. He argued specifically that comparative relation could not be understood apart from number, or, as he put it, "number encompasses all things related comparatively." But when we compare two things we find degrees of equality. That is, it is always possible to find something that is more similar to or more different from an original thing than the thing we first used in our comparison, and so on, ad infinitum. Nicholas concluded that it was therefore "not the case that by means of likenesses a finite intellect can precisely attain the truth about things."

In Book II of his work Nicholas examined what conclusions our reason might lead to regarding the constitution of the cosmos. His results were surprising. Our world could not have a fixed center or a fixed outer boundary, for if it did it would have its own beginning and end. That would be false, because God is the center and the circumference of our world. Reason tells us, then, that the Earth must have motion, that it is a star. Nicholas concluded that the Earth's motion was circular and discussed a host of conclusions in order to demonstrate that humans simply cannot understand such things. The implication was that reason, even when used properly, leads to contradictions that are resolved only in God himself. "We recognize," wrote Nicholas, "through learned ignorance and in accordance with the preceding points, that we cannot know the rationale for any of God's works, but can only marvel."

The clear inference to be drawn from the work of both William of Ockham and Nicholas of Cusa was that the reasoning of philosophy did not lead to demonstrated and unambiguous truth about the world. Both believed that it was important to engage in philosophy in order to make clear its scope. Ockham wished to show that philosophy was not about the nature of real things but about the constructions of the mind. And Nicholas wanted to pursue learning in order to show that in the end it was ignorant. For both scholars, the lesson was that the tenets of revealed truth had to be accepted by faith alone, because reason could not deliver truth.

The Impact of Divine Omnipotence

Theologians and natural philosophers became preoccupied in the fourteenth century with the issue of God's power. This concern resulted less in a change of Aristotelian thought than it did in its reorientation. The increased focus on divine omnipotence affected the way in which scholars regarded the world of nature.

If nature was dependent solely on God's power, then nature could not be thought of as having its own necessary course as it appeared to have, for example, in ancient Greek astronomy. The cause and effect relationships we observe in nature have no necessary connection in themselves. They exist only because an all-powerful God decided that they should go together.

When a rock is thrown into the air, for example, it seeks to return to its natural place in the center of the cosmos, not because Aristotle had identified a necessary property of rocks, but because God had endowed it with that property. The implication was that God could have made rocks to seek a natural place away from the center of the Earth had he wished to. And occasionally, say when God chose to work a miracle, rocks might be made to seek a different natural place. The behavior of rocks was directly dependent on God's will.

Some historians have argued that such a conception undermined the development of natural philosophy because it did not presuppose the unchangeable natural order that had marked Aristotle's thought. But medieval natural philosophers did not believe that God frequently acted capriciously to change usual relationships among things in nature. In fact, they understood that God had created this world by his decree, which they had every expectation was an eternal decree. This understanding meant that God could retain absolute omnipotence and at the same time provide the kind of regularity in nature that would make natural philosophy possible. It was a Christian recasting of Aristotle's system.

Intrigue with Antiquity

Some thinkers began to see in the recovered writings from the past an opportunity to gain access to an ancient wisdom that had been lost. Attempts to restore this learning of old took several forms.

The beginnings of humanism.   Humanism, a major feature of which was the desire to recover and study the works of antiquity, first appeared in fourteenth-century Italy. Humanism's greatest influence on natural philosophy came in the sixteenth century, and will be dealt with in the next chapter. But its beginnings are found already in the fourteenth and fifteenth centuries.

Francesco Petrarca (1304–1374), an Italian known as Petrarch, is often called the father of humanism. Petrarch revived an appreciation for ancient poetry and condemned his own age for letting fall what he called "some of the richest and sweetest fruits that the tree of knowledge has yielded." He was, he said, tempted to regard this ancient learning as more valuable than anything else in the whole world. Petrarch was particularly enamored of the Roman orator Cicero, from whom he learned that communicating knowledge was as important as acquiring learning.

Historian of science Allen Debus has pointed out that Petrarch's love of natural things was instrumental in promoting a new observational study of nature. It also contributed to creating a distrust of medieval scholasticism, based as it was on a dogmatic acceptance of Aristotelian philosophy, which had come to dominate the intellectual atmosphere of the universities. Along with other early humanists, Petrarch made available an alternate outlook, one that would inspire individuals in succeeding generations to explore new intellectual possibilities.

Ancient magic restored. Among the writings of ancient wisdom that began to appear in the Latin West were texts attributed to one Hermes Trismegistus, whose works constitute what is known as the Hermetic corpus. They contained a different view of the relationship of humankind to nature from that common in Western Christianity, one in which nature played a more important role. Hermetic philosophy was a view of nature that held that the soul of the individual could unite with the cosmos. This alternate portrait of nature came into the early modern era largely— although certainly not exclusively—through the efforts of two men, Cosimo de' Medici (1389–1464) and Marsilio Ficino (1433–1499).

Cosimo de' Medici, the first effective ruler in Florence of the famous Medici dynasty that followed him, left as one of his many legacies the translation into Latin of a number of Greek manuscripts he had obtained from Constantinople. Cosimo surrounded himself with artists and scholars, whose company he enjoyed and whose efforts he supported. He was a great lover of books, investing large amounts of the family fortune in establishing a library of illuminated manuscripts and early printed volumes. He also established in Florence a new academy that would focus on bringing to Italy the works of the Greek philosopher Plato. It is believed that Cosimo's physician brought his son, Marsilio Ficino, to the academy, where the Florentine leader immediately recognized the boy's genius and his love of academic study. Cosimo instructed the father to make sure that the lad developed this natural disposition and he subsequently chose Ficino to head his new academy.

Having learned Greek, Ficino undertook Cosimo's mission to translate Plato's works into Latin. His many translations of Plato's works, and his role as head of the Florentine Academy, have solidified Ficino's reputation as one of the central figures in the early modern movement to establish a Platonistic academic tradition as an alternative to the dominant Aristotelianism of the universities.

Cosimo also charged Ficino with translating the works of Hermes Trismegistus, a somewhat mysterious figure thought to have lived at the time of Moses. Ficino was richly rewarded for bringing into Latin Hermes' work, *On Divine Wisdom and the Creation of the World.* His early biographer tells us that he was endowed by Cosimo with the most generous gift of a family estate at Careggi near the outskirts of Florence, as well as with a house in town, and even with beautifully written and costly Greek books by Plato and Plotinus.

The writings of Hermes were later determined to have come from the second or third century A.D., but Ficino believed they were part of a wisdom that predated Christianity. Many themes in the Hermetic writings were similar to those of Christian redemptive history. For example, in both visions it was humankind's main goal to become reunited with God after having fallen from grace. But the tone was different. In the Hermetic view, the fall from grace due to Adam's sin was not regarded as a punishment; rather, it was a necessary phase of the journey to spiritual growth and the ultimate union with the divine. The reunion with the divine did not simply represent a return to what humankind once had. In achieving union with the divine, humankind was transformed into something new that it had never yet been. Hermes drew on many sources outside those common to Christianity, including the religious perspectives of ancient Greece, Babylonia, and Egypt. As a result, Hermeticism embraced more diverse manifestations of the divine than could be found in the Christian understanding.

The place of nature played a much more central role in Hermetic religion than it did in Western Christianity. That was because for Hermes nature was a primary vehicle through which the divine could be known. Ficino accepted that individual souls were part of the world's soul, but that the link between the two had been impaired in the fall from grace. To reestablish the connection required the assistance of the magician—one who had gained an understanding of the secrets that revealed the presence of the divine in the physical world. Through the use of talismans, rituals, formulas, and procedures, the magician was able to get on God's wavelength, so to speak. In this way the magician could know God through nature.

Of course Ficino did not approve of the use of black magic, whose goal was far different from knowledge of the divine. In spite of this disclaimer, there were naturally those who opposed the Hermetic philosophy in the name of a more traditional understanding of God and nature. In fact, Hermeticism did tend to go its own way, producing followers who delighted in resisting the intellectual trends of the day.

Scholars have noted that the relationship of humankind to nature in this Renaissance magical tradition—which has flourished in various forms right down to our own day—carried with it a very different emphasis from that present in Christianity prior to its appearance. In the medieval perspective humans were helpless against nature. But in the Renaissance magical tradition, as a result of a general conviction that spiritual growth could not occur without human effort, the message was that humans should be active and aggressive. Nature was useful, and humans should try to manipulate it for the spiritual benefit it could provide. In this sense the magical tradition of the Renaissance served as a transition between the medieval period and a later era of natural philosophy in which nature's usefulness was exploited for yet different reasons.

The NATURE of SCIENCE

Magic and Science

Using the tools of the magician may seem to be something quite different from the practice of natural science, and of course it is. In appealing to magic phrases and formulae the magician openly cuts short the search for regular causal relationships that link two events and even revels in the mystery connecting them. The fifteenth-century magician would maintain that trying to understand everything that God understood was being too presumptuous and arrogant. If we could merely discover ways to reproduce connections between things that only God was previously aware of, that would be sufficient.

Nevertheless, by striving to uncover these connections, the magician does resemble the natural scientist. The magician wishes to be in control of what happens, even if he or she does not want to know or disclose the connecting links. Just because a modern scientist insists on exposing as many of the intermediate phenomena as possible between two events, that does not mean that the ultimate question of why one event follows another has been accounted for once and for all.

Critics of the notion of causality later on in the early modern era pointed out that saying one thing is caused by another is really saying only that it always follows the other. Humans know how to initiate action to bring about a desired result—to cause the result. To cause something involves an intention to bring it about. We get so used to sequences of events that occur naturally in nature that we say one thing causes another, but in natural science we do not mean that the cause *intended* to produce the result, merely that the effect always follows the cause. A magician does not maintain anything more than that.

Paracelsus. An example of the way the Hermetic philosophy could disrupt established trends in the early modern era can be seen in the work of Theophrastus Phillippus Aureolus Bombastus von Hohenheim, known commonly as Paracelsus (1493–1541). Paracelsus was aware that Hermeticism represented an alternative view to the teachings of the university, particularly in his chosen field of medicine. But he was the type of personality who enjoyed defying authority, especially where the treatment of the sick was concerned.

There is no evidence that Paracelsus ever took a medical degree, although he did attend several universities. He learned some alchemy and medicine from his physician father, who practiced in several mining towns during Paracelsus's youth. But he appears to have gathered the full range of his knowledge primarily from the treatment of patients, particularly as a surgeon accompanying the armies that were continually engaged in warfare.

Paracelsus openly denounced the growing reverence in university medicine for the Greek physician Galen. He rejected Galen's claim that health consisted in maintaining a balance among the body's four humors. He was motivated more by the Hermetic conviction that structural similarities ran through the natural world at all levels. For example, he discarded Galen's belief that opposites cure and replaced it with the principle that like cures like. If the body was affected by a poison, it should be treated by a similar poison. Much to the discontent of the physicians of his day he repeatedly appealed to analogies from chemistry and alchemy in his explanation of health and disease. For example, he believed that the body's organs separated the impure from the pure just as the alchemist did. Illness occurred when the organs failed in their function and impurities accumulated.

Paracelsus produced many followers who continued to champion his iconoclastic approach to medicine in opposition to the Galenists, who wished to restore ancient Greek medical practice. Not all of those who revered the Hermetic philosophy generated as much resistance as did Parcelsus. Still, because of its willingness to incorporate new ideas into its vision, many had regarded the main outlook of Hermeticism with suspicion from the beginning.

A Reformation of Faith

An enduring phase of the recasting of the medieval heritage occurred at the intersection of the intellectual sphere with that of practice. During this period of skepticism some registered their dissatisfaction with the interpretation of Christian doctrine and the way it was put into practice.

Along with the degeneration of the ecclesiastical hierarchy that accompanied the erosion of the authority of Rome came a marked decline in the moral life of the clergy. Such laxity was not entirely new. In different regions of Christendom various unorthodox practices had grown up over time until they were regarded as almost normal. Consequently the moral life that true Christian teaching required had been interpreted in a variety of ways for several centuries prior to the Protestant Reformation of the sixteenth century. As far back as the eleventh century Pope Gregory VII stood firm, at considerable cost, against clergy who had purchased their positions and against clergy who had married. His efforts resulted in his having to leave Rome, ending his days in self-imposed exile.

Not all subsequent popes were as concerned as Gregory VII was to, as he put it, "love justice and hate iniquity." But others in the church periodically felt obligated to denounce corruption, especially when it appeared to have become the order of the day. The efforts of these figures from the fourteenth and fifteenth century prepared the way for the great innovation of the early modern era known as the Protestant Reformation, which produced its own effect on the way in which God's relationship to nature was understood.

Early reformers.   The Oxford theologian John Wycliffe (1324–1384) was a key figure in the call for major reforms in the church. Wycliffe recognized that many church leaders had become indistinguishable from civil authorities in how they led their lives. So enamored were they of the power they wielded that they came to prize the enjoyment of life over their duties and their responsibilities to lead a holy life. Wycliffe's protest against clerics who accumulated considerable amounts of property, and his defense of monastic poverty, clearly put him at odds with many in the church hierarchy who regarded him as an extremist.

The issue that caused the greatest stir as the fourteenth century came to a close, however, was Wycliffe's denial of the doctrine of transubstantiation, the belief that in the sacrament of communion the bread and wine are transformed into the actual body and blood of Christ. Wycliffe did not believe that when Christ said "This is my body" to his disciples at the Last Supper he intended them to take him literally. To be so overly literal took attention away from the meaning of the sacrament, which was to hold in mind Christ's sacrifice and its central importance to a life of faith. Focusing on the material world meant that believers were dwelling on the letter, but not the spirit, of Christ's command to remember him. It promoted superficial, not deep and genuine, spirituality.

Wycliffe's attitude presaged the general view of reformers to come after him about the relative unimportance of the physical world in comparison to the spiritual. For these reformers natural philosophy would not be a major concern since there were much more vital matters. In calling for a return to a purer form of Christianity, Wycliffe was echoing the view of earlier theologians who insisted on subordinating the realm of the physical to that of the spiritual.

The church authorities could not stand idly by while Wycliffe attacked so central a doctrine as transubstantiation. He was condemned on several fronts, especially after he acquired a following among laymen, whom he sent forth to preach his reforms. No doubt he was able to escape personal harm because of the turmoil of the Great Schism that afflicted the papacy around this time.

One of his followers who lived in central Europe, however, was not so fortunate. Jan Hus (1369–1415) attended the University of Prague and was ordained a priest in 1400. Very impressed with the writings of Wycliffe, Hus translated one of them into Czech and made it known widely. He shared Wycliffe's anger at what he regarded as the immoral behavior of many clergy and soon became a thorn in the side of the established ecclesiastical authorities. As the result of a papal bull (decree), the archbishop of Prague decreed in 1410 that Wycliffe's works be burned, an action that Hus denounced from the pulpit. The pope excommunicated Hus in 1411, and in the following year ordered him placed under arrest. The Czech ruler, however, did not carry out the order, and Hus continued to support Wycliffe's views openly. He wrote a work, "Six Errors," which he posted on the walls of a church in Prague and in which authorities claimed to find heretical teachings. Called before the Council of Constance in 1415, Hus attempted to defend his calls for reform. The council, however, condemned him to the stake, and he was executed.

Hus continued to exert an influence on Bohemia after his death. Although different groups of Hussites formed over the course of the fifteenth century, the most lasting formed in 1457 as the Unity of the Brethren, a forerunner of the Moravian Church of the present day. Thus, well before Martin Luther nailed the famous ninety-five theses to the church door in Wittenberg in 1517, Protestants had already appeared in the early modern period.

Erasmus and Luther.   The story of the appearance of Protestantism in the sixteenth century is better known than that of the earlier reformers and it needs only a brief discussion here. Just prior to Luther's action in Wittenberg, Europe had been entertained by the writings of an irreverent scholar who took the name Desiderius Erasmus (1466–1536). Erasmus was born out of wedlock and was forced by guardians to attend a monastery. Although he became a priest, his heart and mind were far from those of a monk. He grew to revere the writers of antiquity, who taught him to bring to his work a secular tone that was detached from official religion. In fact, he delighted in using his eloquent, witty, and sarcastic style against the same kind of excesses and corruption that had motivated earlier reformers, and in favor of what he regarded as true religion.

By the end of the first decade of the century Erasmus's mockery of established religion won him great popularity in Europe. Theologians, he said in his *Praise of Folly* of 1509, are so conceited that they think they are already in heaven, from where they look down with pity on others as so many worms. They protect themselves with a wall of imposing definitions, conclusions, corollaries, and explicit and implicit propositions. Erasmus wished to expose this kind of attitude for what it was so that people could get beyond the letter to the spirit of real religion.

In 1517 Martin Luther challenged his contemporaries with his discontent, expressed in the language of theological argument. He stood in a long tradition of reformers who denounced hypocrisy and pleaded for a return to a more genuine Christian piety. The nineteenth-century Luther scholar Albrecht Ritschl noted that early in his career Luther believed that the issue of salvation was so central that it trumped all other concerns. In other words, both Wycliffe and the young Luther had little interest in claims made about the material world.

In Chapter 4 we shall see that later in life Luther permitted himself to be drawn into consideration of an assertion about the physical world, namely, about whether the cosmos was Earth-centered or sun-centered. Nevertheless, the view held by numerous reformers that the relationship between humankind and God had little or nothing to do with that between God and nature persisted. It would continue to be represented throughout the sixteenth century and beyond, threatening, as one scholar has put it, to sever the vertical dimension between God and nature as if the realm of nature could make no claims at all. Such a view would be invoked in the seventeenth century by those who opposed Galileo's attempt to demonstrate that knowledge of nature could not be ignored when trying to arrive at a true religion.

◎ The Impact of Printing ◎

An early modern innovation of a completely different kind was the invention of movable type in the middle of the fifteenth century. While technological in nature, this change had a profound impact on the development of Western civilization because it helped to spread ideas more rapidly and more extensively than ever before.

The Era of Scribes

From antiquity to the mid-fifteenth century books were produced by scribes, who laboriously copied manuscripts by hand. By the Middle Ages scribes worked independently or as part of a team in a scriptorium attached to a monastery or a records

A Medieval Scribe

office. After the rise of the universities in the Latin West, the so-called pecia system made its appearance in response to the increased pressure for books from university faculties. This system divided the labor so that a copyist worked only on selected portions of a text. Both male and female copyists then received payment for each completed piece from a lay stationer.

The most common material from which manuscripts were made was parchment, made from animal skin. Typically, monks would soak the skins of sheep or goats in a lime bath to remove the flesh, then, while the skins were still wet, they would stretch the skin and scrape it with a knife. Next they would rub it with pumice to raise the nap and with chalk to whiten it. The completion of each technique produced varying qualities of parchment. Once lay stationers instituted the commercial production of manuscripts, the making of parchment became a trade done by parchmenters, whose shops were located in the stationer's town near a supply of water.

Medieval and early modern scribes had to conform to strict rules when copying letters. Some manuscripts used letters of equal height confined between two lines; others used what might be thought of as "lower case" letters. By the Middle Ages copyists used both kinds of writing for most purposes. Sometimes illuminators were employed to add color or images to certain letters of the text. In general, copyists determined the formality of a work by what cut and thickness of pen they used and how slowly and carefully they wrote the letters. In the fourteenth century early humanists like Petrarch, who were very involved in producing books of the works of antiquity, introduced their own reforms into the kinds of scripts used.

Beginning in the thirteenth century paper began to be used for correspondence and around 1400 copyists used it for lower-grade books. Arab copyists had learned how to make paper from the Chinese and employed it as early as the ninth century. Its use spread into the Mediterranean region among merchants for notes and records and by the thirteenth century it was being produced in Italy. Paper made from cotton and linen rags was less expensive to produce, but it did not lessen copyists' time.

Once the pages had been completed, a binder undertook the final stages in the production of a book. Often the sheets of parchment were folded once to form two leaves, or *bifolia*. Writing on the front and back of each half sheet therefore produced four pages. Several of these bifolia were then gathered into *quires*, which were sewn together and attached to wooden covers that were sheathed with leather.

Although the production of books increased after the appearance of Western universities, it was still an extremely laborious process. The notion of identical copies of a book meant something quite different in the era of scribes, when individual differences could not help but exist from one copy to the next, from what it would mean after the advent of printing. Further, the total number of books circulating in society was nothing compared to what there would be once movable type was invented. It has been estimated that fewer books were put out between the early Middle Ages to the middle of the fifteenth century than those printed between then and the end of that century.

The Advent of Printing

The fifteenth century saw the appearance of printed books, a development that brought uniformity to bound volumes that scribes could not duplicate. The initial innovation of printing from blocks of wood was then quickly overtaken by the advent of movable type.

Block printing. Printed books first began appearing in the fifteenth century using a technique that had been employed somewhat earlier to print fabrics and to produce playing cards and devotional images. In block printing a text or an image was carved into a block of wood in such a way that the raised portion of the block formed the letters or picture in mirror image. The block was then inked and paper was placed on the block. Rubbing or pressing the paper on the block transferred the image to the paper.

Initially, the text or image was transferred by printing one page at a time from a single block of wood. In this case printers used but one side of a page and pasted two sides back-to-back before binding. It was not long before presses that could print two sides came along. Woodcut prints, especially of images, could also be added to hand copied manuscripts or they could be printed first, with text added later by a copyist.

The obvious advantage of block printing over copying was that the printer could continue to put out identical copies of a book until the deterioration in the wood blocks affected the quality of the text. But there were disadvantages as well. Each page required its own wood block and preparing the blocks was slow and expensive work. Carving the letters into the wood so that they were well aligned and of even size was extremely difficult. A lot of effort went into the preparation of all the blocks needed for an entire book.

The earliest known European woodcut image dates from the third decade of the fifteenth century. Block-printed books flourished throughout the remainder of the century, but diminished after the invention of movable type. Their subject matter most frequently dealt with biblical stories that could be used by clergy for instructional purposes. For example, a prominent block book, *The Poor Man's Bible,* paralleled events in the life of Jesus to those of the Old Testament in order to portray him as the fulfillment of earlier history.

Movable type. The production of books changed again around the middle of the fifteenth century with the appearance of movable type. The major difference between this approach and that used in block printing was that there was a mold for each letter rather than one for each page. While that meant that the printer needed thousands of letter molds instead of just one mold per page, the molds could be assembled and reassembled to create a long running text much more easily than could be done in block printing. One additional innovation also made a huge difference. The letter molds were made from metal, not wood, so they could be used for a much longer time and they gave a cleaner image.

The story of the invention of movable type in the West is a familiar one. Most everyone has heard the name of Johann Gutenberg (ca. 1400–1468), although his original name was Johannes Gensfleisch. Gutenberg's father belonged to a line of wealthy and powerful merchants in Mainz, Germany. He adopted the name of his estate, Gutenberg, as his family name when Johannes was a young man.

Mainz was an important ecclesiastical principality, whose powerful archbishop had traditionally served as an elector of the Holy Roman Empire. Gutenberg's father and uncle were officials in the archbishop's mint. Knowledge of metalworking had been vital to the church since the early Middle Ages in the making of vessels and icons. It is presumed that Gutenberg learned about casting chambers, the casting of dies, and the stamping of gold coins from this connection to metalworking.

Specific details about Gutenberg's life are sparse. We do know that he moved to Strasbourg around the age of thirty and that he most likely made his living as a craftsman. In the late 1430s Gutenberg became involved in a legal dispute with associates in a business venture. In return for loans, Gutenberg taught his associates various skills, which included the art of printing. What becomes clear from the records is that Gutenberg had been experimenting with an idea he had, but needed outside support to develop it.

Gutenberg's specific idea was a way to cast individual letters from metal that could be used to form text. Others had tried to make metal type, and even to use individual wooden letters, but the problems of perfecting movable type were formidable. It would take an extremely inventive mind to solve the many facets of the problem sufficiently to make the process feasible. The first and one of the most challenging problems was how to cast molds for a typeface that would produce letters of the same height and also accommodate the spacing between such different letters as *i* and *m*. Gutenberg solved this by making two L-shaped pieces that slid into each other and by regulating the height and spacing.

But this was by no means the only problem Gutenberg needed to solve. He had to have a metal that was hard enough to use for printing, but was soft enough to cast. He had to have the right kind of ink. He could not use the water-based ink preferred by scribes and utilized in block printing. An oil-based ink was required. Finally, he had to devise a means for transferring the imprint from the type to the paper. Gutenberg was familiar with presses used in wine and cheese making, and adapted them to his own purpose.

By 1438 Gutenberg was back in Mainz. Having an idea was one thing, but realizing its potential was quite another. Thousand of letters had to be made, which necessitated hiring metalworkers to cast the molds and make the type. Materials had to be purchased, including metal, ink, and paper. He needed people to assemble text in braces, ready for the press. To make the production of a book worthwhile, several presses had to be built and people hired to run them. He made a deal in 1450 with a local businessman named Johann Fust that would enable him to move beyond perfecting his printing technique on small items such as tickets and flyers, and to undertake the huge task of printing the entire Bible.

In 1452 he secured another loan from Fust, on condition that if it was not repaid, he would forfeit the printing equipment. The Bible, once completed, contained two pages per sheet in double columns—some 641 sheets of text. It employed a black Gothic text common to manuscripts of the day and had red and blue leader letters that were added later by an illuminator. It is estimated that less than a few hundred copies were printed, some on vellum and some on paper. To accomplish such an enormous job Gutenberg produced several hundred thousand pieces of type. From a later account we learn that he supposedly used six presses and printed about three hundred sheets per day.

Unfortunately, Gutenberg was unable to repay the loan to Fust on time and he lost his equipment in a lawsuit in 1455, the year the Gutenberg Bible was completed. The sale of the Bible went well, although profits went to Fust, not Gutenberg. In 1465 Gutenberg obtained a pension from a wealthy citizen of Mainz, which supported him until his death three years later.

The impact of Gutenberg's work in perfecting the art of printing with movable type was immediate and profound. There were some who were suspicious of the new technique, assuming that any process that could put out such a magnificent product in so many identical copies came from the devil. But it was not long before printing presses were springing up everywhere. By 1481 they were present

The Gutenberg Bible

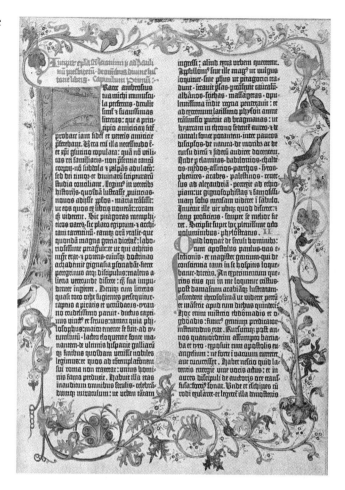

throughout the Holy Roman Empire and by 1500 they could be found in every important municipal center of Europe. The development obviously impacted scribes negatively. The Abbot of Sponheim, for example, exhorted monks not to stop copying just because printing had come along. It was good, he said, to keep idle hands busy, to encourage diligence and devotion, and to promote knowledge of the Scripture. But even he had seen the handwriting on the wall. He had his own works printed, including one entitled *Praise of Scribes*.

As we shall see in Chapter 4, the development of natural philosophy and mathematics benefited greatly from the invention of printing with movable type. Among the many works printed in the sixteenth century, those devoted to astronomy, natural history, anatomy, and other aspects of nature figured very prominently. Printing proved to be a great stimulant to the spread of new ideas about the natural world.

◎ Expanding Geographical Horizons ◎

The lure of travel was not new in the early modern period. The Venetian merchants Nicolo and Maffeo Polo, accompanied by Nicolo's son Marco, had journeyed to China in the thirteenth century, where Marco rose to a high status as a diplomat. In addition to traveling throughout China itself, he went as an envoy to Japan, India, Burma, and Tibet. On returning to Venice after a quarter century's absence, he wrote an account of his travels entitled *The Travels of Marco Polo.* The book became known in Europe and even played a key role in the story of Christopher Columbus.

In the fifteenth century the pace of exploration quickened substantially as travel was no longer left to curious individuals, but undertaken with government support. If the desire to see new and exotic lands was not new, the organization of specifically maritime exploration and the techniques used to carry it out were. What became clear to political leaders of the early modern period was that knowledge of new worlds brought with it an economic advantage. Here was an opportunity for invest- ment that would increase trade with faraway lands that were already known. What happened, of course, was the discovery of lands not already known. This brought with it challenges no one had expected.

The Known World

For nearly everyone at the beginning of the early modern period the world contained four continents—Europe, Africa, Asia, and a vast unknown land in the Southern Hemisphere known as the Antipodes. Contemporary maps show little knowledge of the discoveries of the Norse sailors in the Northern Atlantic; in fact, they show the three known continents more or less crowded together with little space given to the oceans. The center of the world was Jerusalem, with Europe, Africa, and Asia posi- tioned around it in a distortion of the more accurate maps that had been made by ancient Greek mapmakers.

Medieval people knew very little about Asia and Africa. Of course, Christians knew the story of the magi who had come to visit the newly born Christ child from Asia, and Africa was by tradition the land from which King Solomon's vast wealth derived. Africa was also supposedly the home of Prester John, who, according to many, was the apostle John himself. He had escaped death in fulfillment of John 21:22–23, which hinted that he would live until Jesus returned, and now ruled over a vast empire in which there was no crime. In addition to these links to religion, many stories about these regions had been passed down from classical times. Leg- ends of fantasy about rivers of gold, giants, legless birds that never landed, and monsters of various sorts were common.

One of the ancient works that was translated into Latin in the early fifteenth century was Claudius Ptolemy's *Geography.* The translation of Ptolemy's text proved to be influential in important ways. For one thing, Europeans learned new mathematical means of representing three-dimensional surfaces on a two- dimensional plane. This played a role in the development of new techniques in map- making that were already underway. Europeans also learned how Ptolemy himself

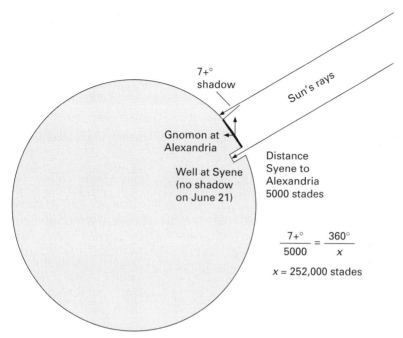

7+°
shadow

Sun's rays

Gnomon at
Alexandria

Well at Syene
(no shadow
on June 21)

Distance
Syene to
Alexandria
5000 stades

$$\frac{7+°}{5000} = \frac{360°}{x}$$

$x = 252{,}000$ stades

Determination of the Earth's Circumference

had depicted the world. Of course, his conception ran counter to the common view that centered the world on Jerusalem, so naturally it elicited opposition and criticism from many church leaders. But for those who regarded the ancient wisdom as authoritative, Ptolemy's understanding carried great weight. They accepted, for example, his depiction of the west coast of Africa at the edge of the map and his assumption that Asia extended a vast distance to the east. This implied that for a spherical Earth, the distance between Asia and Africa was not great.

The determination of the Earth's circumference went back a long time before Ptolemy. Aristotle had calculated its circumference to be about 1.8 times as great as the modern value of 24,901 miles. Eratosthenes in the third century B.C. employed an ingenious means of calculating how far it was around the Earth. He knew of a well, located at Syene (modern-day Aswan, Egypt), where on the summer solstice the sun cast no shadow on the walls at noon. This meant that the well was located near 23½° N latitude, which is the sun's altitude above the equator on June 21. On the same day, at the same time, some considerable distance almost due north in Alexandria, the sun did cast a shadow, the angle of which could be measured. Using the well at Syene as his base point, Eratosthenes determined that if the shadow's angle of a little more than 7° corresponded to the 5000 *stades* of distance along the Earth's surface between Syene and Alexandria, then all the way around the Earth should correspond to 360°. Simple proportionality revealed that distance to be 252,000 *stades*. Exactly how close that measurement is to the modern value depends on how big a *stade* is taken to be, and that is a controversial matter.

Some assert that Eratosthenes came to within 2 percent of the modern value, but an expert historian of early Greek astronomy, D. R. Dicks, has concluded that Eratosthenes' measurement amounts to 29,000 English miles.

Ptolemy himself relied on others for his understanding of how big the Earth was. A century after Eratosthenes, the Stoic philosopher Posidonius calculated it to be 240,000 *stades*, which he later revised downward to 180,000 *stades*. The significance of the latter calculation was that this relatively low figure was the one Ptolemy put into his *Geography*, and therefore the one that became authoritative for key figures of the early modern era.

An Era of Exploration

Starting around 1420 Europeans began taking to the sea in impressive fashion. They opened up new worlds to the imagination, which in their own way contributed to the innovations of the early modern period. To do this they had to rely on techniques of navigation they had learned from the past and they perfected new techniques that would enable them to face the new challenges ahead.

Finding position at sea. Once out of sight of land on a sea that caused the ship to pitch and roll, determining accurate position was not a simple matter. The heavens provided the only means available to calculate latitude, which could be done with some accuracy. Determining longitude precisely required accurate timepieces. The common way to measure time on a ship was the hourglass, which made calculating longitude next to impossible.

To find latitude, or the position north of the equator, it was necessary to determine the altitude above the horizon of a known celestial object such as the Pole Star or the sun. In the case of the Pole Star, which is located above the North Pole, latitude is given directly by the Pole Star's altitude above the horizon. During the day sailors could measure the sun's altitude above the horizon and then find their latitude by consulting charts that showed at what latitude the sun had that altitude for that day of the year.

For both daytime and nighttime determinations of latitude, medieval mariners used a cross staff to measure altitude. The other instrument for measuring altitude available to astronomers, the astrolabe, was too difficult to hold steady enough at sea to give precise results. A cross staff consisted of two pieces of wood, a longer one with lengths calibrated on it, and a shorter one that was held perpendicularly to the first. A mariner would brace himself against the side of the ship, pointing the long stick about midway between the celestial object and the horizon. He would then position the cross piece at the precise spot on the long stick that permitted him to align the two ends of the cross piece in such a way that the sun, for example, was at one end at noon and the horizon was at the other. He could then read the altitude from the calibrations marked on the long stick.

If the latitude and longitude of a destination were known, then a captain could reach it by sailing to its latitude and running along that latitude until he reached his goal. More likely he would determine the course from his starting point and use a compass to reach it. The compass was the most trustworthy instrument at sea. By

A Mariner's Cross Staff

the end of the fourteenth century mariners were using a magnetized, north-pointing needle mounted on a card that had the four main directions marked on it. If the positions of two locations were known, then the compass bearing of one position from the other could be calculated and a ship could follow the course by constantly monitoring the direction traversed. Even when sailing into unknown waters the captain could still maintain knowledge of his direction so that he would be able to return. But determining how far along the bearing the ship had gone in a day was more difficult to estimate, especially when the presence of ocean currents and contrary winds were involved.

Early explorers faced an ominous challenge. Their ships and sails were built such that there was no real hope of maintaining a course in the middle of a major storm. Captains had to lower their sails and allow the storm to carry the ship where it would. More than one discovery of new lands resulted from ships being blown off course. Europeans discovered the Cape Verde islands in the middle of the fifteenth century, for example, while fleeing a storm. Despite the difficulties, mariners became bolder and bolder in their search for precious metals and spices and in their determination to win new converts to Christianity and to bring greater glory to their homelands.

Expanding knowledge of Africa. Africa's west coast provided the first of the continent's shores to be explored. Prince Henry, the third son of the Portuguese king, had distinguished himself in battles on the north coast of Morocco. From traders in the captured region he learned about a gold trade that inspired him to extend Portugal's interest to the south. There was another goal as well. If it was possible to locate a southern entrance into Africa, the Portuguese might be able to find

Prester John and enlist him in the task of removing the Muslims from northern Africa and the Holy Land.

But Prince Henry was not a sailor himself. To send an expedition in this direction along Africa's west coast meant finding a captain who would not be deterred by claims about sea monsters and increasing heat that boiled the ocean. People believed that no one who had ventured beyond the cape at latitude 29° N had ever returned, so they named it Cape Nun. And if Africa were joined to the unknown land of the Antipodes, then it surely could not be circumnavigated.

Much later, in 1625, the claim surfaced that Prince Henry had established a center at Sagres on Portugal's coast in the third decade of the fifteenth century. Soon it became widely accepted that Henry had brought together experts in astronomy, mapmaking, and ship design, and in building nautical instruments that made possible major geographical discoveries. Recent scholarship, however, has discredited this wellestablished and commonly believed tradition, concluding that no such school ever existed and casting doubt on Prince Henry's own abilities in these areas of expertise.

What is clear is that Henry certainly encouraged geographical exploration and that during his lifetime Portugal established bases on islands off Africa's coast. These served as stopping points and supply stations for ships that were departing on or returning from voyages to the south. By 1434 Portuguese captains had ventured some 350 miles beyond Cape Nun. Not long thereafter other voyages returned with African slaves, a cargo that soon opened up a thriving market demand. Just before midcentury Portuguese sailors discovered that the African coast curved to the east, giving them hope that Africa could indeed be circumnavigated. They also gave the lie to the many tales of horror that had kept Europeans out of southern waters prior to Henry's time.

Although the Portuguese were secretive about the details of these voyages, Spain too began to take an interest in the African coast. As a result Portugal persuaded the pope in 1455 to grant them exclusive rights to the regions they had explored so successfully. This agreement did not prevent the Spanish and even the English from seeking their own agreements with the pope regarding various regions of the West African coast or from their engaging in the slave trade. When a new king came to the throne in 1481, the Portuguese determined that they must use their superior experience in the area to establish their claims to the African coast and to secure a short route to India by sailing around Africa.

On the first of several enterprises, the king commissioned a captain to take with him stone markers, which he was to set in place at various conspicuous points along his route. Inscribed in Latin, Arabic, and Portuguese, they laid claim for Portugal. A subsequent voyage in 1487 was to attempt to sail around Africa. The captain, Bartolomeu Dias, went as far south as anyone before him when, without knowing it, he was blown around the southern cape in a violent storm. He then traveled far enough north to convince himself that he had in fact found Africa's southernmost point. On his return home the king named the point the Cape of Good Hope, for a route to India now appeared more likely than it ever had.

The king also sent two overland missions east in 1487, one of which was to try to find Prester John, who was rumored to be near present-day Ethiopia. The other was to go to India via Arabia to find out the routes of Muslim ships that brought spices to the Persian gulf and eastern Africa. The traveler sent to find Prester John died in Cairo, and the other adventurer, Pero da Covilhã, also never made it home. But word of what Da Covilhã had discovered did reach the king. He had made it to the western coast of India and then traveled far enough down the eastern coast of Africa to suspect that the Indian Ocean did in fact join the Atlantic. This word, when it reached the Portuguese monarch, sustained him and his successor in their continuing quest to find a sea passage to India.

Having learned from the experience of others before him, Vasco da Gama (ca. 1469–1524) embarked in 1497 on a voyage supported by the Portuguese crown that was to succeed in reaching India. After rounding the Cape of Good Hope, he continued north to the port at Mozambique, where he encountered Arabs instead of the African chiefs that had usually greeted the ship. One of those on board who could speak Arabic learned that there were great ports to the north and also that Prester John held many cities along the coast, although John himself supposedly lived far inland from the shore. In one of these ports Da Gama found Hindus, whom he believed to be Christians, confirming in his mind that Prester John was close by.

With the help of a local pilot, Da Gama crossed the Indian Ocean to the west coast of India, where he was not well received. His fleet of ships represented a new threat—different from the occasional European traveler—to the established trade that Arabs and Persians conducted with India. But now that the sea route to India had been uncovered, the Europeans would not be turned back. News of Portugal's accomplishment spread far and wide in Europe after Da Gama's return in September of 1499. Although the Portuguese kept the details of the voyage secret, their success inspired other countries to set out on their own.

"Discovering" the new world.   For Europeans, discovering a sea route to India confirmed a supposition that numerous scholars had harbored for some time. They knew that India existed even if they were uncertain that Africa could be circumnavigated. It was a matter of discovering whether or not a route could be found.

Christopher Columbus (1451–1506) had a similar goal. He wanted to find a sea route to China by sailing west. He had studied *The Travels of Marco Polo* and had consulted Ptolemy's *Geography*. From these sources he concluded that Asia extended extremely far to the east. If that was so, then the distance China's far eastern shore lay from the west coast of Africa, on an Earth of the small circumference Ptolemy reported it to be, could not be immense. The work of the Italian physician and geographer Paolo Toscanelli (1397–1482) confirmed that the distance was not great. Toscanelli had determined that the distance from Europe to China by sea was some 5,000 nautical miles. Using Toscanelli's computations selectively, Columbus was able to reduce the estimate to 3,500 miles, still an impressive distance to span, to be sure.

The novelty of Columbus's proposal, of course, was that he would sail west into the unknown. Hugging the coast of Africa was relatively easy in comparison. Other than the Azores, which were known to the Portuguese, there was no known land far out in the Atlantic. Columbus became more and more convinced of the feasibility of his plan, but persuading sponsors for the trip proved to be very difficult. As early as the 1480s he began trying to visit the courts of Portugal and Spain to present his case. The Portuguese eventually turned him down, and the Spanish throne only came around to his plan after long debate.

As is well known, Columbus left Spain for the Canary Islands in August of 1492. He intended to sail west along the latitude line of the Canaries because he knew there were favorable winds there at that time of year. But Columbus did not rely on celestial navigation to maintain his course. Born in Genoa, Italy, his earliest maritime experience had occurred in the Mediterranean, which is essentially all at the same latitude. To navigate there sailors had learned a technique known in English as "dead reckoning," a phrase derived from "deduced reckoning" and abbreviated as DED reckoning. It relied on the technique referred to earlier using a compass and determining the distance traversed. From a beginning point a navigator set the course with a compass and then measured the distance traveled each day, marking the new position with a pin on a chart.

Maintaining the direction by compass was one thing, but calculating the distance traveled was more difficult to do with precision. To accomplish it the navigator determined the ship's speed and then multiplied it by the time traveled at that speed. Speed and distance were measured every hour by throwing a piece of flotsam overboard and observing when it passed by a mark on the ship's rail. At that moment the pilot began a chant, stopping when the flotsam passed a second mark on the rail. After noting the syllable where the chant ended, the pilot converted the length of the chant into a measure of speed, thus permitting calculation of the distance made in that hour. In Columbus's case, he deliberately underreported the distances traversed in order not to alarm the crew more than necessary about how far they were from home. When he finally made landfall in the Bahamas in October, Columbus was convinced from the gold ornaments worn by the inhabitants that he had in fact reached his destination of Asia. On his return voyage he composed a letter to his Spanish royal sponsors announcing his success in reaching islands off the coast of Cathay that were rich in gold and spices and whose inhabitants could be converted to Christianity. Subsequent voyages to the region did not persuade Columbus that he was mistaken about having reached Asia, but there were others who became convinced that the lands reached were neither China nor Japan.

Among these was the Italian geographer Amerigo Vespucci (ca. 1450–1512), who visited what is now the north coast of Brazil in 1499. Two years later he followed up on a finding of the Portuguese explorer Pedro Alvares Cabral, who had reached the east coast of Brazil in 1500. Vespucci traveled south along the coast and became convinced that the land he saw comprised a continent not known before to Europeans. In 1504 he was quoted as saying that these lands across the ocean could rightfully be called a new world. When a German publisher put out a new map in 1507 in which these regions were depicted, he used Amerigo's name to label the new continent under the mistaken impression that he, not Columbus, had been the first to observe it.

By the beginning of the sixteenth century, then, the world was a different place from the one it had previously been for Europeans. In their self-centered and chauvinistic manner, they regarded the journeys of Da Gama, Columbus, and others as voyages of discovery, as though the people who lived in these regions represented something wholly new to existence. That the inhabitants of the so-called New World had their own history and civilization, which should be cherished for what it was, never occurred to them. Their goal for the future of the western lands was in one way similar to their goal for the East; they desired to make new converts to Christianity. But it was different as well. The task was not to engage the New World in trade, as it had been with China and India, but to conquer and exploit these previously unknown territories for the wealth they could bring to European monarchs.

There is a certain irony about the status of natural philosophy in particular and learning in general at the turn of the sixteenth century. In one sense the Western world was filled with the innovations of the early modern era that have been discussed in this chapter. New religious and intellectual ideas, a new technology of printing that permitted ideas to spread rapidly, and new realities about the Earth's geography captured the imagination of everyone but those in lowest classes, for whom the business of survival occupied all their attention. The world was turning out to be a different place from the one that most people had known. The future seemed to call for a new outlook.

On the other hand, especially where natural philosophers were concerned, scholars were increasingly looking backward, not forward. They strove to revive and regain a wisdom from the past. As we shall see, this striving to regain a lost wisdom was about to produce an even greater reorientation of humankind's view of itself and of its place in the cosmos than that produced by the innovations we have just examined.

Suggestions for Reading

Elizabeth Eisenstein, *The Printing Revolution in Early Modern Europe*, 2nd ed. (Cambridge: Cambridge University Press, 2005).

John R. Hale, *Age of Exploration* (New York: Time, Inc., 1975).

David C. Lindberg, *The Beginnings of Western Science* (Chicago: University of Chicago Press, 1992).

CHAPTER 4

—————————◎—————————

The Renaissance of Natural Knowledge

The meaning of *renaissance,* when applied to the early modern period, is "rebirth," or "beginning again." Evidence of a renaissance of European society in the period from the fifteenth through the seventeenth centuries consists of a number of developments. For example, commerce, industry, and cultural life expanded as the number and role of middle class burghers—residents of the town—increased and altered older feudal relations. Among the educated elites, new intellectual institutions emerged to offer an alternative to those predominant in the medieval world. And on the level of religious culture, there appeared a new secular outlook that challenged the dominance of the clergy's conception of learning.

These changes occurred first in Italy as the republican city-states began to give way to powerful leaders who wished to consolidate their authority in central governments. Everywhere in Italy the values of rebirth were evident. But they did not remain confined to Italy alone. Especially as the sixteenth century dawned, the notion of new beginnings spread to the countries of central and northern Europe.

◎ Humanism ◎

One particular expression of this rebirth of European society was named *humanism* by scholars in the nineteenth century. It is of interest to us as far as the natural sciences of anatomy, natural history, and astronomy are concerned. Humanism concerns the life of the mind, in particular, the acquisition of knowledge and the interpretation of its meaning. The fundamental assumption of this aspect of the Renaissance was that humans in the past had crafted a wisdom that had been lost, should be restored, and could serve as the foundation on which to build a new outlook. This ancient wisdom had been compiled by thinkers of old whose writings had been lost or corrupted and whose original texts were just now beginning to be recovered. Because many of these texts were pre-Christian, humanists exhibited a secular interest that took its cue not from a transcendent Christian God, but from the human individual. The humanist scholars of the Renaissance integrated this secular outlook into their understanding of Christianity.

There were points where humanism and the Christian attitudes that had settled into the Latin West clashed. Humanists valued individual intellectual exploration of what the mind could discover. This presented a problem as humanists began to occupy chairs at universities, because the prevailing scholastic attitude was to defer to dogma, especially the established philosophical and theological interpretations that were based on the teachings of Aristotle. At Oxford University, for example, one statute decreed "that Bachelors and Masters who did not follow Aristotle faithfully were liable to a fine of five shillings for every point of divergence." If humanists valued individualism, the scholastics certainly did not.

Because they integrated Christianity into a wider perspective, humanists did not automatically regard the course of history as an unfolding of God's plan. That was too linear and deterministic for them. They believed that humans contributed to the formation of history, that God had given humans abilities to use in shaping history. Consequently, they believed that history could ebb and flow, so to speak; it did not proceed linearly, but was cyclic.

Humanists were impressed with knowledge that was useful and they often criticized the scholastics for using their rational gifts to pursue goals that were too esoteric to be practical. When it became clear that the ancients had possessed useful knowledge and wisdom that had been lost, it was important to humanists that the contents of ancient texts be recovered and mastered to learn what they contained. For them, to "begin again" meant to study ancient texts in order to regain a perspective that had been abandoned, or perhaps distorted, and to evaluate it using the reason of the present. This attitude of humanism, we shall find, became the predominant motive of some important sixteenth-century figures who turned their attention to the natural world.

◎ The Study of the Human Body ◎

Restoring what the ancients knew about the human body and using that knowledge as a basis for beginning the study of anatomy anew is a story that took a particular turn over the course of the first half of the sixteenth century. It involved breaking away from received tradition by returning to the anatomical knowledge of the ancient world as a starting point for a new anatomy.

The Heritage of Anatomical Knowledge

It is sometimes thought that the reason anatomical knowledge was in such an undeveloped state during the Middle Ages was that the church forbade the dissection of humans. The church did forbid boiling bodies to obtain skeletons, a practice crusaders sometimes employed in order to obtain bones that could be transported home for burial. But by the fourteenth century there was no prohibition against, for example, postmortem dissection for the purpose of investigating the cause of death. That did not mean, however, that anatomy was a subject eagerly pursued for its own sake. As a result, knowledge of the anatomy of the human body was not well developed at the beginning of the sixteenth century.

Mondino's *Anatomy*.   One source that did exist had been written in 1316 by a professor at Bologna named Mondino dei Luzzi (ca. 1270–ca. 1326). He had read the only source that was available to him from the ancient Greek physician Galen. It was not the physiological treatise whose later recovery would exert a powerful effect on the study of anatomy; rather, it was a short version of Galen's work that had come into Latin from the Arabic. It was not helpful at all in setting a tone for how to study anatomy. In fact, it dealt simply with digestive organs and the limbs, but its use marked an early indication of interest in the Greek physician.

Mondino wrote up an outline of a procedure for dissectors that became something of a standard text in medical studies. Well into the fifteenth century most universities required medical students to observe one or two dissections a year, often prescribing Mondino's *Anatomy* as a text. By the late-fourteenth century Galen's work, *On the Use of the Parts,* did appear in translation, but it was relatively ignored. Mondino's work had become the standard text, especially after 1476, when the first printed version of Mondino made its appearance. It was reprinted several times on into the sixteenth century.

By 1400, then, anatomical dissections were conducted in medical courses of many universities, with others adding the practice over the course of the century. A professor would read from Mondino's *Anatomy* while a demonstrator, surrounded by student onlookers, pointed to the various organs and parts the professor described. Commentaries on Mondino's book also began to appear, written by professors who lectured on anatomy.

The recovery of Galen.   By the beginning of the sixteenth century the study of anatomy was beginning to be regarded more highly than it had been. Now the enthusiastic examination of ancient texts brought new translations of Galen's works directly from the Greek. After 1500 Galen's *On the Use of Parts* was available and in 1523 a translation of his treatise on physiology, *On the Natural Faculties,* appeared. Then in 1531 a newly discovered text by Galen, *On Anatomical Procedures,* was translated into Latin with a commentary by a medical humanist at the University of Paris named Johannes Guinther (1487–1574).

With Guinther's textbook in 1536, *Anatomical Institutions According to the Opinions of Galen for Students of Medicine,* Galen's authority over Mondino and scholastic anatomy was established. With the recovery of Galen came changes from the way anatomy had been studied earlier. He began with the skeleton, not well known to Mondino, and then proceeded to the muscles, nerves, arteries, and veins of the arms, hands, and legs. Next came the internal organs, divided according to function. Such a systematic plan was something quite new and it soon generated enthusiasm to the point of hero worship.

Because of the general prohibition of human dissection at the time of Galen, he had carried out anatomical dissection on animals, trying to make reasonable inferences to human anatomy. This practice naturally resulted in some unavoidable errors, which were exposed by the practitioners of human dissection in the Renaissance. These two occurrences—the general enthusiasm for Galen's authority and the recognition of his mistakes—ran in opposite directions and created a challenge for the anatomists of the mid-sixteenth century.

Andreas Vesalius's On the Fabric of the Human Body

One of Guinther's students was Andreas Vesalius (1514–1564), a native of Brussels. He began his medical education at Louvain, then went to Paris, where he assisted Guinther in the preparation of the latter's textbook. From Paris he went to Padua, the site of the most famous medical school in Europe. After obtaining his doctoral degree in 1537, he conducted public human dissections for medical students and others at the university as a lecturer in anatomy and surgery.

During the years he lectured at Padua, Vesalius introduced a number of innovations into anatomical instruction. First, he did the dissections himself. He did not rely on a demonstrator to do them, as was the earlier practice. Further, he performed the dissections more often than had been done before and he even gave private dissections for more advanced students that went into greater detail than those done in public. He made charts to help the students master the complex detail they encountered as they watched him dissect, and he published six of them separately in 1538. Finally, he also dissected animals, not to make inferences to human anatomy, but to compare them to humans and show similarities and differences.

During these years Vesalius began work on his own anatomy textbook. While he appreciated his former teacher Guinther's devotion to ancient texts, Vesalius did not respect his abilities as a dissector. Further, Vesalius had worked out his own solution to the tension that existed between respect for Galen's authority and the results he himself uncovered as he dissected human cadavers. He retained his reverence for Galen, but at the same time he wished to make what he regarded as necessary corrections to Galen's work on anatomy.

In writing his book, *On the Fabric of the Human Body,* Vesalius made clear his humanist motivations. That is, he wished to restore the ancient wisdom and at the same time to build upon it. In his dedicatory preface to the Holy Roman Emperor, Vesalius noted that although anatomy in some universities was beginning to approach the importance it once had enjoyed in ancient times, the knowledge of medicine was still woefully inferior to that of long ago. Vesalius wished to change that, he said, by restoring the status of anatomy from the low level to which it had degenerated over time to the position it once held.

To do this, Vesalius set out to restore the kind of medicine Galen had practiced. He determined that he would honor Galen's excellence by holding him up as a model. In his book he followed the order now becoming popular among Galen enthusiasts—he began with the bones, not the soft internal organs of the body.

Vesalius also wanted to imitate Galen's conception of himself as a philosopher. He spoke of anatomy as a branch of natural philosophy, a branch that he wanted to recall from the dead. He did not propose to undertake a theoretical revision of Galen's philosophical stance; rather, he chose to expand upon it. As a result, his work not only provided a vastly greater amount of anatomical detail than had Mondino's brief descriptions, it even increased the coverage given by the master himself.

Wherever Vesalius's dissections uncovered errors in Galen, he did not hesitate to point them out. But Galen's errors in no way undermined Vesalius's goal of restoring ancient knowledge. He explained the errors by noting that Galen had had to rely on animal dissection. The mistakes had come about because Galen had not adopted the views of ancient physicians who predated him and who had based their conclusions on human dissection.

For all the emphasis on human dissection, that was not the only unusual aspect of Vesalius's book. More immediately striking was his use of anatomical illustration. Although some medieval manuscripts contained hand-painted pictures of the human body, illustrating the text was not commonly done. Mondino's *Anatomy,* for example, did not contain any anatomical illustrations. After the invention of printing a few anatomical texts contained woodcuts, but Vesalius's new work was by far the most heavily illustrated. He tied the illustrations so closely to the text that their use became indispensable.

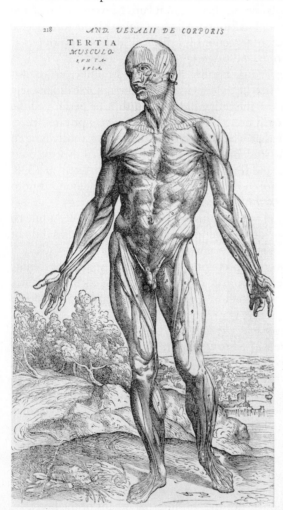

Vesalius never identified the authors of the illustrations even though their collaboration was essential to the purpose of the book. The images were works of art in their own right, revealing a depth and clarity of detail that was truly amazing. Many preserved a sense that it was a living individual who was being depicted. The illustrations were more impressive than those of any earlier anatomical text, and they were only outdone by the anatomical

Vesalius's Anatomical Illustration

illustrations Leonardo da Vinci had completed at the end of the fifteenth century but never published.

On the Fabric of the Human Body was intended for practicing physicians, not medical students. It contained much more information than a medical student could use, nor could students generally afford to purchase such a large and well-illustrated volume. Vesalius published a shorter version for them in the same year the larger book appeared. Called the *Epitome,* it was an outline of the *Fabric* printed in a larger format with wall charts and oversize anatomical figures.

Among those who were highly impressed with what Vesalius had produced was the emperor to whom it was dedicated. Charles V of Austria, Holy Roman Emperor, quickly appointed Vesalius as physician to the imperial court. Vesalius did not hesitate to leave university life in order to accept a position of royal patronage. As we shall see again in other contexts, natural philosophers who valued freedom from the regimented duties of teaching were generally happy to accept the support of a patron.

Not everyone welcomed Vesalius's work, however. Strict Galenists, for example, reacted against his correcting the master so glibly. But many correctly recognized that here was one of those rare works that would endure. A second edition appeared in 1555, consistent with the enormous influence the book was to have on its time. The work stood as a testimony to the greatness of past works that had both inspired it and provided the model on which its innovations were based.

◎   Natural History: Animals, Vegetables, Minerals   ◎

In 1555 there appeared in London a translation from the Latin of a book by Peter Martyr, entitled *The Decades of the Newe Worlde.* It was about the discoveries of the Spanish in the New World. "The diligent reader," Martyr observed, "may . . . learne many secreates touchynge the lande, the sea, and the starres." In the course of his narrative, Martyr referred to the ancient author Pliny, who, he reported, "hathe wrytten in xxxvii. bookes al that perteyneth to the naturall historie."

Martyr was using the phrase "natural history" in its original very broad sense, as information about the natural objects, plants, and animals of a certain place. Natural history, then, referred not just to the animal world, as it tends to in modern usage, but to all natural objects, animal, vegetable, and mineral. In their treatment of these subjects, Renaissance figures once again used ancient authorities as their point of departure.

The Kingdom of Animals

As is evident from Martyr's reference, the Renaissance took its cue from Pliny the Elder (ca. 23–79), who had written a massive work on natural history in the first century. Pliny was a Roman nobleman, who, after a military career, had become a prolific writer on a myriad of subjects.

Where animals were concerned, Pliny's descriptions formed the basis for so-called bestiaries in the Middle Ages, which were also available to humanist writers of the Renaissance. These were books on animals in which symbolic meanings were equally as important as defining characteristics. Following on Pliny's own practice, bestiaries included references to monsters and marvels, including fabulous creatures like dragons and unicorns. The Old Testament of the Bible, as well as ancient pagan sources, provided precedents for the use of animals as symbols of larger truths.

In the middle of the sixteenth century the Swiss humanist Conrad Gesner (1516–1565) composed a massive natural history. Gesner, who studied ancient languages and medicine in Paris, finished his doctorate in 1541 and lectured on natural philosophy in Zürich until he was appointed a municipal physician there in 1554. His five-volume *History of Animals* began appearing in 1551 and ran to some 4,500 pages!

Gesner did not hesitate to include information about monsters and imaginary animals, but his division of the animal kingdom in keeping with Aristotle showed his respect for ancient Greek sources. While he was dependent on the natural historians who preceded him, Gesner did on occasion add his own observations to the work. His *History of Animals* became well known, especially after notable artists provided illustrations to add to those Gesner had included from earlier authors.

Plants and Their Uses

The growth of humanism encouraged the study and depiction of plants, just as it had the study of animals. Latin and Greek translations of a work by the ancient Greek Theophrastus, called *History of Plants,* were available in the 1490s. Sixteenth-century

Illustration of a Sea Satyr from Gesner's *History of Animals*

humanists, however, were not as interested in this description of the parts and classification of plants as they were in the more practical use of plants and herbs as medicines.

Here the ancient authority was Dioscorides (fl. 50–70), a Greek physician who served in Nero's army. Dioscorides had a particular interest in medicines and in his travels he sought out pharmaceutical remedies of all sorts. He authored a five-volume medical work that was translated into Latin in the sixth century as *De Materia Medica.* Volume 1, on plant materials, contained descriptions of the store of plants familiar to the ancients, which Dioscorides claimed to have used in his own practice.

As was the case with both Galen and Vesalius, a shorter summary of Dioscorides became more popular than the longer work. Dioscorides' herbal, *Ex Herbis Femininis,* proved to be more useful to physicians of the Middle Ages. It contained but seventy-one plant medicines, none of which was unfamiliar to Europeans, as many of the ancient plants in the larger work were. Among the Latin encyclopedists there were those who added their own descriptions of plant medicines to the ancient literature. By the late fifteenth and early sixteenth centuries, then, there was a long tradition of interest in herbal medicines. Many of the books of herbals contained illustrations that had been copied from earlier manuscripts, some of which were based on ancient paintings of plants in Dioscorides.

When the humanist scholars of the Renaissance went back to Dioscorides, they tried to produce as accurate a text as they could. They brought to their work the assumption that also motivated Vesalius; namely, they wished to supplement the ancient wisdom with the knowledge they had from their own experience. In their first preparations of the early Greek texts, they simply left illustrations out. Further, in their quest for the pure text, they corrected errors from the original that had crept into later editions of Dioscorides. In 1544 Pietro Mattioli (1501–1577), famous in Italy for providing the first mention of the tomato, provided an edition with commentary on the *De Materia Medica* that was reproduced many times, guaranteeing that Dioscorides would remain the most popular authority on herbal medicine in the sixteenth century. Mattioli did include illustrations, including those of new plants unknown to Dioscorides.

The German school of botany. In the 1530s and 1540s several important botanists appeared in Germany who changed the way herbals appeared. Above all they inserted illustrations of plants—prepared by skilled artists—that were based on their own observation.

The first of these German botanists was Otto Brunfels (ca. 1488–1534), an early convert from Catholicism to Protestantism. After serving unsuccessfully as a pastor in two different German communities, he went to Strasburg in 1524. Here he established a school and turned to the study of natural science and medicine. His medical interests took him into botany, which Brunfels believed suffered greatly for want of adequate illustrations. His goal was to produce a work in which the illustrations were of plants as they actually appeared, not illustrations that resembled an ideal natural state. He took this so seriously that he directed an artist-illustrator to depict the plants just as they were—including broken leaves and whatever effects insect pests might have produced. His work, entitled *Living Portraits of Plants According to the Imitation of Nature,* appeared in three volumes between 1530 and 1536.

Leonhard Fuchs (1501–1566) continued the tradition of correcting the errors in earlier editions of Dioscorides. Like Brunfels, he was raised Catholic, but converted to Protestantism. He studied medicine at Tübingen, became a professor there, and in 1542 wrote *History of Plants.* Fuchs listed the plants in his book alphabetically by name. His illustrations were also his own, not those of Dioscorides on whom his work was based. They were done by artists, who tried to depict the plants realistically in some 516 pages of illustrations. While only the first illustration—of the *Anemone sylvestris*—was in full color, many were given in a single color other than black.

Finally, Hieronymus Bock (1498–1554), another early Protestant pastor who combined the study of medicine with his training in theology, served early in his career as garden inspector for the prince. In 1533 he met Brunfels, who, impressed with the work Bock shared with him, encouraged Bock to publish. In 1539 Bock's new herbal appeared in two parts, neither of which contained illustrations. This put Bock at a disadvantage, especially when compared to the massive work of Fuchs in 1542. Therefore, a second edition published in 1546 contained some 546 illustrations, followed in 1551 by a third edition with 70 more. These illustrations did not match the outstanding quality of those Fuchs had published, but Bock's work demonstrated the trend toward accurate illustration that had become the norm. What made Bock's work notable was its outstanding textual description of each plant's habitat and especially its medicinal effects, a feature lacking in Fuchs's work. Together these three German scholars—Brunfels, Fuchs, and Bock—helped to create a new era in herbal literature.

Herbaria and Botanical Gardens.   Other influences on the history of botany that proved to be of great importance were the invention of the herbarium and the establishment of botanical gardens. Herbaria were collections of plants that had been dried, and were preserved by pressing them between sheets of paper. The first known collection of this kind was created in Italy at the University of Bologna by Luca Ghini (1490–1556). Ghini was a member of the faculty of medicine there until 1544, when he was brought to the University of Pisa by the grand duke of Tuscany. By the time he died a dozen years later, herbaria were commonly found all over Europe. Because plants changed with the seasons and were not always available, herbaria provided a means of preserving them for teaching purposes or to serve as models for draftsmen to make illustrations for publication. Another indication of the flourishing of botany was the establishment of professorships of botany in medical schools of the time. The first such position was created at the venerable medical faculty of Padua in 1533 and then by at least five other universities over the course of the sixteenth century. At Padua the course of study initially involved mastering Dioscorides. With the passage of time, however, the new trend to supplement ancient wisdom with new information based on personal observation had an impact on medical education.

The new professors of botany soon found themselves responsible for an additional duty—the administration of botanical gardens. The presence of herb gardens in cloisters was an old tradition, going back at least to the late Middle Ages. The herbs supplied spices for the kitchen and medicines for common ailments. Now, the formal cultivation of herbs acquired a distinctly new status. Leonhard Fuchs had complained in 1542 that physicians were still ignorant of plants, but his work—along with that of the others we have discussed—changed that substantially. By the end of the century there were botanical gardens at Pisa, Florence, Bologna, Paris, and Montpellier. Not only did these gardens afford a means by which medical schools could further medical knowledge and education; they also demonstrated to the larger public that plants had acquired a significance that went well beyond their aesthetic value.

Understanding Fossils

The word *fossil* originally meant, simply, something that had been dug up out of the ground. When scholars wrote about fossils they included—besides what would be classified as fossils today—substances such as mineral ores, metals, crystals, and useful rocks. This wide array of objects, then, contained some items that looked like organisms and many that had no similarity whatever to living things. When there was a similarity it was certainly noted. Leonardo da Vinci (1452–1519), for example, held that the well-preserved fossil shells of northern Italy were so similar in general appearance to the mollusks living in his day in the Mediterranean Sea that they had to be organic in origin.

But this was the exception. Most scholars did not believe that even these fossils had once actually been alive. For one thing, often the fossils were of organisms that were extinct, so there was no existing living organisms with which to compare them. For another, depending on how they were preserved, many fossil remains of organic creatures were hard to interpret, and even remain so today for modern paleontologists.

Also, there were alternative explanations of fossils that made clear to humanist scholars why they were shaped as they were. When the shapes of forms resembled organisms or stars or some other object in the cosmos, the natural philosophers who were impressed with Plato's thought explained the similarity as a manifestation of the unity of the cosmos. A similarity born of the basic structure of reality linked the two objects, but it in no way required that one originated from the other. An Aristotelian natural philosopher, on the other hand, might say that just as simple organisms were formed by spontaneous generation from nonliving materials, a similar process might have occurred in the depths of the earth or in the sea to produce fossils that resembled simple organisms. In both Platonic and Aristotelian explanations fossils that resembled living things did not come from living things.

Agricola on fossils. Georg Bauer (1494–1555), better known as Agricola, wrote *On the Nature of Fossils* in 1546. He began his studies with the Latin and Greek classics before he decided to study medicine. After taking his degree at the University of Padua in 1526, he settled into a life that included medical practice, politics, and especially the study of mining and minerals. In the same year that he published his book on fossils, he was elected mayor of the Saxon town of Chemnitz, where mining was a local industry with a long tradition.

In his book, Agricola summed up what a long list of ancient authors had said about fossils. Again we see the typical attitude of the humanist—respect for the ancients, but a willingness to correct them when self-acquired knowledge contradicted them. Agricola changed the way knowledge about fossils was organized. Previously, they had been listed alphabetically by name, an approach Agricola found much too arbitrary. He proceeded to organize fossils by their physical properties. For example, gems were all put into one group, as were rocks, metals, hardened fluids (salt, amber), and stones (gypsum, mica).

Gesner's *On Fossil Objects*.   In addition to the massive writing on the history of animals described earlier in this chapter, Conrad Gesner also studied objects that had been dug up from the earth. He knew and respected the work of Agricola, but he set out to produce a larger work on fossils. If the *History of Animals* is any indication, the anticipated project would have dwarfed everything that preceded it. The larger work never appeared, however, because of the author's untimely death in 1565. But he did publish a volume in that year that he intended as a preliminary book to the bigger project. Its full title was *A Book on Fossil Objects, chiefly Stones and Gems, their Shapes and Appearances*.

Reverence for classical antiquity was what motivated Gesner to give to his planned venture an encyclopedic scope. He gave the objects he was describing in his book their Latin and Greek names in addition to German labels. He wanted to restore as much as possible what the ancient authors had said and then add to that what he had learned in the present. This attitude was consistent with his sympathies, as a Swiss Protestant, for the Reformation's rejection of authority in favor of one's own experience. The reformers wished not to simply accept what church authorities said about the Bible, but to read it for themselves. Gesner learned Hebrew as well as Greek in order to read for himself the ancient texts of the Bible, rather than versions that had been corrupted as they were passed down.

As in his work on the history of animals, and in the works of Vesalius in anatomy, Gesner insisted on illustrating the text. A well-known historian of geology has noted that Gesner's work was the first in which illustrations were used systematically to supplement a text on fossils. This provided a means of checking that the names being used for fossils denoted the same objects that others meant when they used those names. In Agricola's work, for example, the absence of illustrations often left one in doubt about exactly which object the author was describing.

By using illustrations as he did, Gesner also confirmed how important firsthand experience was to him. He compiled his material as much as possible from his own observation. To prepare the illustrations, he employed experts who could do the job, but he insisted that they work under his supervision to guarantee a satisfactory degree of loyal representation.

Gesner's work on fossil objects was also important because it drew on another development of the sixteenth century—the formation of collections of specimens. As we saw earlier, botany led the way, through the new herbaria, in the deliberate

Gesner's Woodcut of a
Living Crab and Its Fossil

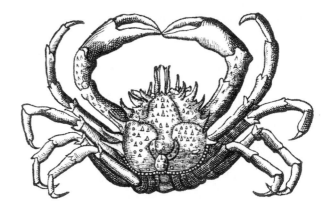

Pagurus la-
pideus, parte
ſupina expreſ-
ſus.
Ein ſteininer
Meerkrebß/o-
der Täſchen-
krebß.

formation and preservation of collections. Gesner helped to utilize collections for the study of fossils by referring to an existing collection. He had the catalog of the collection of a friend, the physician Johann Kentmann (1518–1574), bound together with his own book. On opening the work the reader encountered first Kentmann's catalog, whose frontispiece was an illustration depicting a cabinet with twenty-six numbered drawers. Alongside the cabinet was a key, entitled "The Ark of Fossil Objects of Johann Kentmann," that explained what was in each drawer. Of course, some of the drawers correlated with objects that would not be classified as fossils today. For example, drawer 10 referred to marbles and drawer 9 referred to gems. What moderns consider fossils were found in drawer 5 (stones) and drawer 12 (wood embodied in rocks).

In his work Gesner thanked his friend Kentmann for supplying him with specimens of objects he did not have. His endorsement of the catalog meant that other catalogs would soon appear, marking the increasing presence of museum collections of fossil objects. Historian Martin Rudwick has observed that "without the establishment of a tradition of museum preservation, it is difficult to imagine how a science of paleontology could have emerged."

Gesner's book was also the first to call for cooperative research on fossils. He explicitly stated that he had written the work to create interest in the field and to obtain from readers examples of stones that would enhance the collection depicted in his volume. Gesner knew that fossils varied a great deal from one region to the next, so that the study of fossil objects required cooperative efforts of many natural philosophers. Such a network formed a scholarly community, sustained by correspondence. Formal scientific societies would come in the next century; nevertheless, Gesner's work provided a stimulus in the direction of cooperative research.

For Gesner a major motivation for studying fossils, and in his other work on animals, was to bring greater glory to God by discovering the divine wisdom behind the Creation. In the case of fossil objects, Gesner pointed out in the dedication of his book that the gems he described were worthy symbols of the building materials of the heavenly City of God.

One way to bring glory to the Creator was to show nature's usefulness. Gesner was a physician by training and vocation. He promised in this preliminary book that he would, in the larger volume on fossils to come, describe the power and nature of every kind of stone and mineral. Like most investigators of his time, the prospect of an object's powers seamlessly incorporated several characteristics at once: their usefulness, their ability to reflect the macrocosm in the microcosm, and their general status as part of God's creation. Gesner helped to impart a greater significance to the study of fossils than it had had before.

◎ The World of Copernicus ◎

A new system of astronomy appeared in the middle of the sixteenth century that immediately caught the interest of European astronomers and eventually captured the attention of scholars in general. The cosmos was about to grow much, much larger.

The Humanist Background of Sixteenth-Century Astronomy

Like other disciplines, astronomy felt the effects of humanism. As we shall see, two important astronomers from central Europe—Regiomontanus and Copernicus—were steeped in the works of the ancients. Copernicus, the one from Poland, would leave the greater impact on Western natural philosophy.

Regiomontanus. Astronomy and mathematics initially felt the impact of humanism most significantly in the work of Johann Müller (1436–1476), a fifteenth-century Bavarian scholar who was known as Regiomontanus. He was the son of a miller, but his keen intellect and his abilities in mathematics provided him with a means of rising above his low social origins. At 11 years of age he was already attending the University of Vienna, where he completed his bachelor's and master's degrees as rapidly as university regulations allowed.

The university was known for its cultivation of astronomy and cosmology and Regiomontanus quickly took to these subjects. Along with one of his professors at the university, he became involved in a project to make Ptolemy's *Almagest* more accessible than the original. Regiomontanus, who knew Greek, brought the project to completion in the early 1460s under the title *Epitome of the Almagest*. He supervised the various dimensions of the undertaking, updating Ptolemy's work through the addition of later astronomical observations and revised computations, and critical reflections on the text.

In the preface of *Epitome of the Almagest,* Regiomontanus openly exhibited his sympathies for humanism. His profound reverence for antiquity showed through clearly in his celebration of the marvelous achievements of the ancient Greek mathematicians. By contrast, he noted the woeful state of the study of mathematics in his own day. He took advantage of the newly invented technology of printing to produce multiple copies of numerous ancient mathematical, astronomical, and geographical texts. His *Epitome,* however, did not appear in print in Rome until 1496. In the fall of that year a young Polish scholar, also the product of a humanist education, traveled to Italy, where his studies would include astronomy.

Copernicus's education.   The budding Polish scholar was Nicolaus Copernicus (1473–1543), youngest child of a merchant in the town of Torun. Young Nicolaus must have shown promise in his early education because his uncle decided that he should follow in his own footsteps as a churchman and scholar. In 1491 Nicolaus went to Cracow to study at the university, although there is no evidence to suggest that he completed a degree there. Founded in the middle of the fourteenth century, Cracow was one of the universities in central Europe that was under the influence of humanism as it moved north from its birthplace in Italy. Here Copernicus mastered Latin and the classics and here he began the study of ancient mathematics and astronomy.

In 1496 his uncle, who had by this time become a bishop, sent Copernicus to Italy to study church law, as he had. In Bologna, where the oldest European university had been founded in 1088, humanism reigned supreme. Although Copernicus was there to master canon law, his interest in astronomy and mathematics continued unabated. He was able to learn from Domenico Maria da Novara, a professor of astronomy, who passed on to Copernicus his enthusiasm for ancient Greek astronomical works.

By his early thirties Copernicus had learned Greek, earned his degree in canon law, studied medicine, and returned to Poland, where he became an assistant and attending physician to his uncle, the bishop. His uncle had been successful in obtaining a position for him as canon in the cathedral at Frombork while Copernicus was still in Bologna. This provided a regular income, but did not require specific clerical duties of him.

Astronomy in Crisis

For several years Copernicus lived with his uncle in the bishop's residence in Lidzbark Palace, not far from Frombork. During these years Copernicus came up with the first version of his plan for improving Ptolemy's account of the heavens. Ptolemy's system needed improvement for at least two reasons, one theoretical and the other practical. First and foremost, Ptolemy had resorted to a technique that seemed to Copernicus to violate the most fundamental sense of a rational explanation of heavenly motions as required by the Platonic tradition, which Copernicus had embraced through his humanist education. An explanation of the heavens that lacked rationality was not the kind of world God had created. Copernicus was determined to replace that technique with an alternative that met the criteria of rational explanation. Second, Ptolemy's system was encountering practical difficulties when it was employed in the service of society.

Ptolemy's "irrational" system.   Among the assumptions of the ancient Greeks was that motion in the heavens shared the quality of perfection that characterized the entire realm above the orb of the moon. Heavenly bodies moved because they were carried on rotating celestial spheres, which, recall from Chapter 1, were assumed to be physically real. So the paths traced out by celestial bodies were perfect circles and the rotating movements of the spheres were uniform. To permit the path of a planet, for example, to deviate from the circle, or to speed up or slow down, introduced into the system an element of unpredictability, of imperfection, of irrationality.

But all observers of the heavens knew that the planets went faster and slower at different points in their orbits. How was this nonuniform motion to be explained? The Greek astronomers had constructed combinations of interlocked spheres, each one turning with uniform circular motion, which, when taken together, produced the observed irregularities. If we could be transported to the planet itself, we would not feel accelerated or decelerated motion; rather, we would experience only the uniform circular motion of the sphere carrying the planet itself. So the irregularities that were perceived in the heavenly motions from the vantage point of Earth were merely apparent. It was impossible, as Copernicus put it later, "that a heavenly body should be moved irregularly by a single sphere." Irregular motion could only result from combinations of uniformly rotating spheres.

The problem was that the predominant system of astronomy of the day, that of Ptolemy, did permit actual irregular motion in planets as they moved around the Earth. In order to improve the fit between what was observed and what the system depicted, Ptolemy had introduced a device, called the equant, which resulted in an actual acceleration and deceleration of planets as they moved at different points in their orbits (see Chapter 1). To Copernicus that meant that the moving power driving the system was not constant, but fickle, and that was a prospect before which "the mind shudders." To Copernicus and other humanist astronomers of his day who venerated the divine rationality of the ancient world, Ptolemy's equant was an inconsistency of the most fundamental kind. It represented the most glaring deficiency of astronomy and it had to be removed. For this reason, scholars have emphasized that Copernicus viewed his work in astronomy as a *restoration* of pure ancient astronomy.

Practical problems.   A theoretical challenge of this kind was not the only thing wrong with astronomy in Copernicus's day. There were practical problems as well. These practical problems stemmed from the increasing discrepancy between the predicted position of celestial objects and the positions where they were observed to be. The basic reason for this increasing lack of fit was that the passage of time had magnified any original discrepancy present in the system at the time of Ptolemy into much more glaring discrepancies by the late-fifteenth and early-sixteenth centuries. Ptolemy had accounted for the naked-eye observations of his era as best he could. But the fit had not been perfect and in the intervening millennium, as the heavens had continued to turn, this original tolerable lack of fit had grown to intolerable dimensions.

As a result there were—besides the *Almagest* itself—numerous systems that were based on Ptolemy's work but that had tried to improve it by altering the specific devices it contained. There was not just one Ptolemaic system, devised by the master himself; on the contrary, there were numerous "Ptolemaic" systems. For example, one system might try an epicycle where Ptolemy had used an eccentric circle, another altered the rates at which the interlocked spheres turned, and some had even returned to the older system of Aristotle, which used only homocentric spheres.

But none of this revision of Ptolemy, or return to Aristotle, worked to solve the crisis in astronomy. None of these "Ptolemaic" systems gave results that accurately matched up with the naked-eye observations of the day. This was especially obvious because the scholars of Copernicus's day had accumulated many more observations than those that had come down from antiquity. That made the errors more glaring than ever. There was no longer any denying that Ptolemy's system needed improvement.

One area in particular felt the practical effect of astronomy's increasing inability to depict the heavens accurately. For some time, the papacy had been aware of a growing problem with the calendar. It was important for the church because the holy day of Easter was determined by a formula that involved the annual motion of the sun and the monthly motion of the moon. (Easter is the first Sunday following the first full moon after the spring equinox.) The problem was that there were no common units with which to measure the length of a day and the time of the sun's complete trip through the zodiac.

Julius Caesar had decreed that a year was 365¼ days long, meaning that every fourth year there would be an extra day added to the normal 365. But the solar year is a bit more than eleven minutes longer than this, meaning that over time the accumulating minutes of the solar year were being overlooked. Keeping to the formula meant simply ignoring the extra time as the months of the year passed, so that by the end of the 365 days the accumulated extra time still remained. To just continue in this way had the effect of backing the calendar up because the new calendar year began before the solar year was over.

By Copernicus's time some ten extra days had accumulated. That meant that New Year's Day was celebrated ten days prior to when it should have occurred. This became particularly noticeable when the spring equinox, when every place on Earth has twelve hours of daylight and twelve of darkness, came on March 11 instead of on March 21, when it was supposed to occur. Because the celebration of Easter was determined by the spring equinox, the church was rightly concerned.

Regiomontanus was consulted on calendar reform by Pope Sixtus IV in 1475. A few decades later Pope Julius II sought the advice of astronomers and scholars on the same issue. One of those he consulted was Nicolaus Copernicus, who by 1514 had gained a reputation as an astronomer. This reputation was in part due to a work he had completed and circulated privately. It was his first attempt to improve the state into which Ptolemaic astronomy had fallen.

The Achievement of Copernicus

We have only a few works from the pen of Copernicus, and not all of them are on astronomy. Here we will consider two works that he wrote about the movements of the heavens. We will also look at another work written by a disciple who attempted to summarize the achievement of his master. Taken together, these three works gave Western civilization materials that literally changed the world.

Early effort.   Sometime before 1514 Copernicus completed a small work that has come to be known as the *Commentariolus,* or *Little Commentary.* It contains the basic ideas of his later, more developed, and more famous book. In the very first paragraph he makes clear what his major motive was—to follow the ancient astronomers in the wisdom of their assumption about the principle of regularity. Copernicus was exhibiting a conservative, as opposed to revolutionary, tendency when he agreed with the ancient Greeks, who thought that a heavenly body always moved with uniform velocity in a perfect circle and that by combining such regular motions in various ways they could account for the apparent nonuniform motion of the planets.

Copernicus noted that the earliest astronomers introduced eccentric circles and epicycles when simple concentric circles would not give a good enough account of the heavens. When Ptolemy came along, however, his account was not adequate without the introduction of certain equant points. That meant that the planets moved with nonuniform motion. For this reason Copernicus wrote that Ptolemy's system "seemed neither sufficiently absolute not sufficiently pleasing to the mind."

In listing the assumptions of his new system, Copernicus quickly served notice to the reader that his purification of ancient astronomy would be creative indeed. Among his unconventional assumptions he included the following:

a. The Earth is the center of heaviness for objects and it is the center of the moon's orbit, but it is not the center of the cosmos.
b. The sphere of the fixed stars is so enormously far away from the Earth that the distance from the Earth to the sun is imperceptible by comparison.
c. The daily revolution of the stars around the Earth is caused by the Earth's turning on its axis.
d. The annual motion of the sun is caused by the Earth's motion of revolution around the sun, which it shares with the other planets.

The *Little Commentary* proceeded to provide a few details about the apparent motions of the sun; the moon's motion; the motions of the superior planets Saturn, Jupiter, and Mars; and the movements of the inferior planets Venus and Mercury. Copernicus explained how he could replace the equant point by using a circle that was concentric around the sun with two small epicycles. By this means he duplicated the improved accuracy of the equant without making the planet actually move irregularly. Thirteenth- and fourteenth-century Islamic astronomers had used the exact same device, although they had not used it in conjunction with a heliocentric cosmos. So removing the equant was not an improvement that automatically entailed putting the Earth in motion. This would prove to be important to later astronomers as they studied Copernicus's new system.

Since the *Little Commentary* was not printed, it circulated only among a limited number of scholars. Not long after it was written, the pope requested Copernicus's advice on calendar reform. Copernicus declined, saying that until the motions of the sun and the moon were better understood, the problem could not be solved. Clearly he wanted to finish the larger work to which he alluded in his short introductory piece. He would spend the next two decades perfecting the longer work.

Georg Joachim Rheticus's summary. Four years before Copernicus's death a young lecturer from Wittenberg, Germany, came to Frombork. He had earned his master's degree in 1536 and was recruited to teach lower mathematics and astronomy. The young man's father, a physician, had been convicted of swindling and was beheaded when Georg was a teenager. He felt it best to change his family name, Iserin, to dissociate himself from the shame this had brought on the family. He became known as Rheticus, derived from the region in which he lived.

In 1538, Rheticus (1514–1574) decided to take a leave from Wittenberg, probably because of his association with a radical poet who had been expelled from the community. He went to Nuremberg, where Regiomontanus had lived earlier. There he met a publisher of astrological and astronomical manuscripts who had heard about the ideas of Copernicus, most likely from those who had seen Copernicus's *Little Commentary*. Such radical notions appealed to Rheticus, and he determined that he would go to Frombork to learn from Copernicus himself. He took with him as gifts bound volumes of astronomical work from Nuremberg to demonstrate that these well-known German printers would be able to do a good job with anything Copernicus might want to send back for publication.

At this early stage in the Reformation, being a Protestant did not prevent Rheticus from remaining with Copernicus until 1541, when he returned to Wittenberg. In his isolated position at Frombork, Copernicus must have enjoyed having someone present with whom he could share the technical achievement of his study of the heavens. For his part, Rheticus urged Copernicus to have the longer work he had been preparing published. But Copernicus was unwilling because he wanted to incorporate into it some of the new information that was contained in the materials Rheticus had brought him. He did, however, give Rheticus permission to write a summary of his ideas, which Rheticus sent back to Nuremberg in the fall of 1539. Rheticus returned to Wittenberg in 1541 for a short time, before taking up duties at the University of Leipzig.

The summary Rheticus wrote was called *Narratio Prima*, or *First Report*. It contained a general account of what Copernicus had done, with sufficient detail that it whetted the appetites of the astronomers of his time. Rheticus began by identifying his teacher as an astronomer comparable to Ptolemy because he had completed an entire system of astronomy. He noted that because Ptolemy had relied on the observations of his own day, and because an unnoticed error at the foundation was greatly increased by the passage of time, Copernicus had had to build astronomy anew rather than merely correct it.

Rheticus delighted to point out that his teacher was able to liberate astronomers from the equant through an ingenious use of eccentrics and epicycles. He announced that Copernicus accounted for observations of the planets' proper motions by having the sun occupy the center of the cosmos while the Earth revolved, instead of the sun. He gave six reasons why the old assumptions of ancient astronomers had to be relinquished. Number six was that Ptolemy had not paid attention to the need for his system to be internally harmonious. There was no necessary agreement among the parts, as if a musician need only tune one string to produce harmony, without insisting that the others also be adjusted. This was not a problem for Copernicus because the sun governed the celestial motions, so their harmony was under its control.

This appeal to the Greek philosophical belief in the complete rationality of nature resonated with humanist astronomers everywhere. Rheticus's seventy-page account was printed in 1540. It aroused sufficient interest to produce a second printing in 1541.

Rheticus was moving beyond a simple critique of Ptolemy's compromise of nature's rationality with his equant, to a more general indictment: Ptolemy's system was internally inconsistent. He did not use one geometry of the heavens to explain the positions of the planets; rather, he had employed seven different arrangements of spheres to capture the motions of the seven planets of the ancient world. But the arrangement he had used for Venus, for example, had nothing to do with the one employed for Mars, nor was it linked to the arrangement used for any of the other planets. It was a theme that Rheticus had learned from Copernicus himself, one that was to be repeated in Copernicus's own masterwork.

On the Revolution of the Heavenly Orbs.   Rheticus finally persuaded Copernicus to have his work published in Nuremberg and he carried a copy of the manuscript with him when he returned to Wittenberg in 1541. He took a leave from Wittenberg in the spring of 1542 to bring the manuscript to Nuremberg and see it through the printing process. In the fall he took up a new position in Leipzig, and the process of proofreading the text was turned over to Andreas Osiander, a Protestant clergyman who was interested in mathematics. The book, *On the Revolution of the Heavenly Orbs,* came out in 1543.

When he received a copy of the work in April, Rheticus was shocked to note that his role in producing the work was nowhere indicated. Worse than that, in the front of the work was an anonymous preface, later determined to have been inserted by Osiander, that purported to explain away the central idea of Copernicus's work—that the Earth moved whereas the sun was at rest in the center of the cosmos. The author of the preface explained that Copernicus had introduced this idea merely to make calculations and that he was not saying that the Earth actually moved. If this was not understood, the preface concluded, then the reader would "depart from this study a greater fool than when he entered it." Rheticus was incensed and, in three copies he gave others as gifts, he marked out the preface with red ink.

It is clear that Rheticus was angry because he knew, as the one astronomer closer to Copernicus than any other, that his teacher did believe that the Earth actually moved. In his own preface, written as a dedication to Pope Paul III, Copernicus acknowledged that the idea that the Earth experienced movement would seem absurd to astronomers. He then confessed that he had considered communicating his ideas, "written to prove the Earth's motion," only to a few people rather than

Osiander and the Motion of the Earth

The NATURE of SCIENCE

Osiander's position about the motion of the Earth has been cited as an early example of a philosophical position known as antirealism. In this view, scientific theories should be regarded as assumptions that are adopted because of their usefulness, not because they represent nature "as it really is." Theologians in the sixteenth century were familiar with the employment of hypotheses, particularly as a means of trying to ferret out heresies. That is, they would assume a particular proposition and then reason from it to see if any of the conclusions they came to through logical deduction turned out to be heretical. If so, then they could establish that the assumption was theologically problematic.

In adopting a proposition, then, they accepted it only provisionally. They in no way embraced the proposition's truth. Osiander adapted this use of hypotheses to the situation Copernicus faced. Suspecting that the assumption of the Earth's motion would in fact lead to heresy, he wanted to make sure that the reader did not embrace the proposition of the Earth's motion as true. In Osiander's eyes Copernicus had his own reason—mathematical, not theological—for adopting the assumption of the Earth's motion. Osiander acknowledged that mathematicians, too, could use hypotheses profitably, but his main concern was to remove all suspicion of heresy from the work of Copernicus.

publishing them. From this and other similar phrases it is clear that the Earth's motion was not just a hypothesis he had adopted to improve his calculations.

In sentiments similar to those Rheticus had signaled in his *First Report,* Copernicus complained to the pope about the many different systems that tried to render an account of the heavens. They used different assumptions and, referring to the equant, even permitted the fundamental principle of the uniformity of motion to be violated. In a famous passage, Copernicus recorded his distaste for the lack of harmony among the parts of the various Ptolemaic astronomical systems in use. "With them it is as though an artist were to gather the hands, feet, head, and other members for his images from diverse models, each part excellently drawn, but not related to a single body, and since they in no way match each other, the result would be a monster rather than man."

When he got down to business, Copernicus reviewed the reasons that the ancients had believed the Earth was at rest in the center of the cosmos and then argued why these reasons were insufficient. Anticipating the question, for example, of how the clouds could remain suspended in air if the Earth beneath them were moving at a tremendous speed, Copernicus argued that the air was associated with the Earth and would move with it.

Later in the work Copernicus explained how, by endowing the Earth with separate motions, he could account for what we see on a daily and annual basis. Allowing the Earth to turn once on its axis in a twenty-four-hour period produced the same visual effect of turning the entire heavens around a central Earth in the same period. Further, permitting the Earth to travel around the sun once a year accounted for other

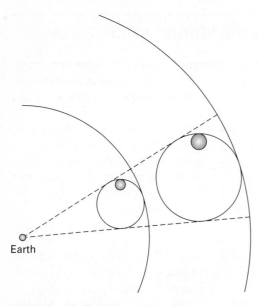

The Ptolemaic System

observations. As the Earth changed its position from month to month, the sun appeared against the background of the different constellations of the zodiac over the course of the year. And the Earth's annual motion also naturally explained why planets like Mars occasionally underwent retrograde motion. As the faster-moving Earth caught up with and passed its neighboring planet, Mars appeared to slow up, stop, go backward, stop again, and then resume its previous course.

Already in the *Little Commentary* Copernicus had realized that putting the Earth into motion around the sun meant that the size of the cosmos had to increase. Now, in *On the Revolution of the Heavenly Orbs,* the implications became even clearer. It now became possible to determine the order of the planets.

In Ptolemy's system a planet's distance from Earth could not be determined because the planet turned on an epicycle. Was the planet carried on a small epiclic sphere close to the Earth, or farther from the Earth on a larger epicycle? In both positions the observational lines of sight from Earth would be identical, so it was impossible to know how far away the planet was. That meant that the order of the planets could not be nailed down. Was Mars closer or farther away than Jupiter? Was Venus on this side of the sun or on the other side?

This was not the case in the new Copernican cosmos. There the order was determined by the geometry of the heliocentric, or sun-centered, cosmos; in fact, the relative sizes of the orbits could be calculated precisely. Consider, for example, the simple case of an inferior planet such as Venus, whose circle of orbit around the sun

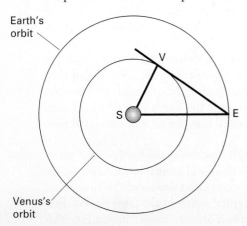

The Copernican System

is contained in the larger circle of the Earth's orbit. (The demonstration for superior planets is slightly more complicated.) The angle SEV is known from observation; it is the angle of Venus's maximum displacement from the sun. The angle SVE is also known; it is a right angle formed by a tangent drawn to a circle from an external point that intersects the radius of the circle at the point of tangency. And, if two angles of a right triangle are known, then all three are known. That means that we know the proportions of the sides of the triangle. So we know the proportion that SV is of SE;

that is, we know the proportion Venus's distance from the sun is of Earth's distance from the sun. This proof gives relative sizes of the orbit, not the absolute distances, but relative size is all that is needed to determine the order of the planets.

With the relative sizes of the orbits set, the general size of the cosmos had to increase. Copernicus pointed out that the farther a superior planet was to Earth, the smaller and less frequent was its retrograde motion as the Earth caught up to it and passed it. So Saturn underwent retrograde motion less frequently and to a lesser extent than did Jupiter, and Jupiter less than Mars. But the stars showed no shifting at all! So the gap between Saturn and the sphere of the fixed stars must be truly enormous. The Copernican cosmos was vastly larger than it had been in traditional cosmology.

Once he had chosen the sun as the center of the cosmos, these implications about the order of the planets and the size followed from observations and from the properties of mathematics. But why had he chosen the sun to be the center in the first place? Here Copernicus's humanism showed through clearly. After setting the order of all the planets, he justified placing the sun in their middle through an appeal to ideas he had absorbed while studying ancient texts. It was an appeal to ancient philosophical traditions whose wisdom carried religious overtones that were assigned to the sun itself: "In the middle of all sits the Sun enthroned. In this most beautiful temple could we place this luminary in any better position from which he can illuminate the whole at once? He is rightly called the Lamp, the Mind, the Ruler of the Universe; Hermes Trismegistus names him the Visible God, Sophocles' Electra calls him the All-seeing. So the Sun sits as upon a royal throne ruling his children the planets which circle round him."

Copernicus's book fell onto fertile ground. There were numerous astronomers who appreciated the magnitude of the crisis astronomy faced and who were as enamored of ancient philosophical perspectives as was Copernicus. An old and well-known claim that Copernicus had written a book nobody read has been shown to be wrong. Thanks to the persistent efforts of a few modern scholars in tracking down and examining copies of the first and second editions of the book, it has been possible to determine a great deal about the reception of the work. There was a rich tradition in the sixteenth century of writing in the margins of a work by those who read it and that has enabled scholars to fill in a great many of the gaps in our knowledge of the events the book set in motion. These events were as profound as any in the history of science.

Suggestions for Reading

Marie Boas-Hall, *The Scientific Renaissance, 1450–1630* (New York: Dover Publications, 1994).

Owen Gingerich, *The Book Nobody Read: Chasing the Revolutions of Nicolaus Copernicus* (New York: Walker and Co., 2004).

Thomas Kuhn, *The Copernican Revolution* (Cambridge: Harvard University Press, 1992).

Martin Rudwick, *The Meaning of Fossils: Episodes in the History of Paleontology* (Chicago: University of Chicago Press, 1985).

CHAPTER 5

―――――――――◎―――――――――

Heliocentrism Considered

In spite of the preface added to Copernicus's book that set aside any claim of the Earth's actual motion, Copernicus himself did believe that the Earth moved. It is abundantly clear that the person who had worked closest with Copernicus, Georg Joachim Rheticus, understood this. Rheticus was extremely unhappy with Andreas Osiander, who soon became known as the author of the anonymous prefatory comments.

By focusing on the issue of the Earth's actual motion, however, Osiander had struck a nerve. For most, this question was certainly incidental to the main task of the astronomer. That task was to depict the motions of the heavens in a mathematical description that was internally consistent and as accurate as possible. As Osiander put it, the astronomer "will adopt whatever suppositions enable the motions to be computed correctly from the principles of geometry for the future as well as for the past." It was a purely mathematical exercise that had only the most minimal concern with the physical state of the cosmos. In trying to make sure that the reader did not think Copernicus was overstepping the boundaries of mathematics into a consideration of physics, Osiander inadvertently drew particular attention to the Earth's motion.

But the question of how one took the hypothesis of the Earth's mobility continued to linger among astronomers in the sixteenth century. Most scholars initially dealt with it by trying to figure out a way to remove it from Copernicus's work, leaving the rest of his system intact. They had good reasons, both theological and physical, for dismissing the prospect of the Earth's actual motion as simply irresponsible and absurd. A few took the Earth's actual motion seriously, however. Copernicus's student, Rheticus, accepted it just as his master had done. And later, at the end of the century, Johannes Kepler incorporated the Earth's motion into his system. The consideration of heliocentrism in the sixteenth century, then, included most who rejected and a few who accepted the Earth's motion.

◎ The Immediate Reception ◎ of Copernican Heliocentrism

Among those who reacted to the publication of Copernicus's book, *On the Revolutions of the Heavenly Orbs*, the interest of scholars who taught astronomy was to be expected. Those in the university town of Wittenberg, heart of the Protestant Reformation, established a particular attitude toward what Copernicus had done. But the work also caught the eye of a foremost theologian of the day from Wittenberg, the Reformation leader Martin Luther, who also had definite views about the book.

The Response Within the Religious Community

Martin Luther nailed his ninety-five theses on the castle church door in the town of Wittenberg in the fall of 1517. He had been teaching at Wittenberg University since 1508 and he remained there as he continued to defy Pope Leo X. Students, faculty, and others in Wittenberg rallied around him, and by 1527 the new Lutheran doctrine had become permanently established there.

By 1539 Luther had heard about Copernicus, presumably from a faculty colleague. In that year, before either Copernicus's early manuscript or his book had been published, Luther made a general comment in the course of one of his "Table Talks" about a new astrologer who wanted to prove that the Earth, not the sky, moved. One of those present wrote down Luther's reaction. He had dismissed the notion, the writer noted, because it turned the whole of astronomy upside down. Luther attributed the idea's appearance to a trend in which everybody was just trying to be clever. For his part, Luther believed the scripture, in which Joshua had sought the Lord's help in making the sun, not the Earth, stand still.

But if the reformers provided no sustained measure of the impact of Copernicus (there is no evidence that the other major Reformation figure, John Calvin, had even heard of Copernicus), neither did the official Catholic Church. In 1545 Pope Paul III, to whom Copernicus had dedicated his book two years earlier, convened a council whose main purpose was to counter the spread of Protestantism. The Council of Trent would last until 1563 and its primary action was to undertake reforms to tighten up what had become lax behavior among clergy at all levels. Nowhere in the long time it existed did the Council consider the book by Copernicus.

We do know, however, that a theologian from the papal court criticized Copernicus in an unpublished analysis of the work that he wrote a year after Copernicus's book appeared. His criticism centered on Copernicus's ignorance of physics and theology; otherwise, the author contended, Copernicus would never have permitted himself to entertain a hypothesis that all philosophers and theologians knew to be contradictory to sound physics and scriptural interpretation. As was the case in the Protestant reaction, there was no attempt to evaluate the technical mathematical merits of the Copernican theory.

The Wittenberg Interpretation

Copernicus's book did receive heavy scrutiny from a group of faculty members at Wittenberg University. There emerged at this newly formed university a consensus on how to read Copernicus's book that has been called the Wittenberg interpretation. Among its features, two stand out. The first was its enthusiastic endorsement of Copernicus's removal of the equant, the off-center point Ptolemy had used that resulted in planets undergoing actual accelerations and decelerations. Like Copernicus, the Wittenberg astronomers felt that heavenly bodies could only undergo uniform circular motion. To them, permitting a planet to experience actual acceleration introduced into the heavens an element of irrationality and unreliability that the presence of uniformly moving circular orbits avoided. The heavens were perfectly rational and predictable; hence any perceived acceleration was merely apparent, the result of combinations of uniform circular motions.

A second feature of the Wittenberg interpretation was the wish to transplant the means Copernicus had used to remove the equant into a system that did not require a moving Earth. If for the Wittenberg scholars Copernicus's great achievement had been to show that it was possible to remove the horrid equant, then—given the unfortunate difficulties that were caused by hypothetically setting the Earth into motion—the obvious goal was to find a system in which they too removed the equant without adopting Copernicus's embarrassing hypothesis of a moving Earth.

Led by Philipp Melanchthon (1497–1560), Luther's articulate defender, these men all believed that the Creator had ordained the study of mathematics and astronomy, and that through such study they could gain unique insights into God's nature. Many of the numerous students of astronomy at Wittenberg later became astronomy professors in universities throughout Germany. Where Copernicus was concerned, Melanchthon appreciated the technical accomplishments of Copernican heliocentrism, but he himself did not focus on them. Rather, he emphasized the need to get back to a geocentric system so that there would be no contradiction between an alleged terrestrial motion and the scriptures. But his influence in supporting and encouraging others in Germany who were trying to preserve Copernicus's achievement while restoring geocentrism was truly remarkable.

Among those who did take up the technical challenge was a Wittenberg astronomer named Erasmus Reinhold (1511–1553). Reinhold openly stated that Copernicus stood as an equal to Ptolemy and that he had restored astronomy from the ruins into which it had fallen. Reinhold remained neutral on the question of the Earth's motion. The annotations he added to his copy of *On the Revolution of the Heavenly Orbs* came in the later technical sections of the book where individual planetary motions were described without using the equant. So convinced was he that these descriptions were more accurate than others that he constructed a new set of astronomical tables using them. These Prutenic Tables quickly became known as a compendium of more accurate data. One other person must be mentioned here, a scholar we met in Chapter 4. Georg Joachim Rheticus came to Wittenberg in 1532 from southwestern Austria. He caught Melanchthon's attention and, on finishing his master's degree in 1536, was appointed as a lecturer in lower mathematics. But he left Wittenberg in 1538 for Nuremberg, where he heard about Copernicus. His role in bringing Copernicus to the attention of the Wittenberg scholars and in getting Copernicus's work published was told in Chapter 4. His return to Wittenberg as a professor in 1541 was brief; the following year he moved on to Leipzig.

Rheticus is important here because he exhibited a different representation of the Wittenberg interpretation. Yes, he was convinced that one of Copernicus's great achievements was to have eliminated the equant from planetary astronomy. But he did not see the need to try to devise a means for transferring Copernicus's techniques for getting rid of the equant into a geocentric system. He was content, as was Copernicus himself, simply to accept a heliocentric system in which the Earth moved.

Rheticus even wrote up a tract, printed much later, to show how one was to understand passages in the Bible that seemed to run counter to Copernican theory, including the one in which the sun was made to stand still. However, the majority of the Wittenberg scholars wished to avoid having to reinterpret scripture by creating a geocentric system that was somehow equivalent to the system of Copernicus. That, in fact, is exactly what occurred in the work of a remarkable astronomer from Denmark, who himself visited Wittenberg as a young man.

◎ Tycho Brahe and the Copernican Theory ◎

Some natural philosophers who studied astronomy took up the challenge to keep the advantages Copernicus had introduced into astronomy while at the same time retaining an unmovable Earth. One of the most important was a Danish nobleman named Tycho Brahe (1546–1601). He made numerous contributions to the development of astronomy, not all of which emerged from his concern with the system of Copernicus.

The Denmark of Tycho's Day

For over a century prior to the sixteenth century the three lands of Denmark, Sweden, and Norway were joined together under the Danish king in order to resist the growing threat of German encroachment into the region. They especially needed to counter the influence of Germans from various cities who had joined together into what became known as the Hanseatic League. The intention of the members of this league was to promote a German monopoly in the lucrative fishing trade around the Baltic Sea.

By the 1520s, however, tensions had built up sufficiently between Denmark and Sweden that the union was no longer tenable. Although Norway remained united with Denmark, Sweden broke away and became an independent land in 1523. Still, the Danish state in the sixteenth century extended beyond the borders of present-day Denmark. It included parts of what is now southern Sweden and northern Germany.

One effect of the German involvement in Scandinavia was the spread of Protestantism. Christian III, who succeeded his father as king in 1534, had been a duke in Schleswig, a Danish province with many German inhabitants. His teachers included German Lutheran reformers and as a young man of eighteen he was present at the Diet of Worms, where Luther took his stand against the church authorities. He did his best to introduce Lutheran Protestantism into Schleswig and as duke made no secret of his wish that Denmark would become Lutheran. In 1536, as King Christian III, he declared Denmark Protestant, thereby bringing Lutheranism to Denmark as a whole.

Against this background of change the first of Denmark's famous natural philosophers, Tycho Brahe, made his name. His father and uncle were both members of the Rigsraad, which consisted of about twenty of the country's most influential nobles. Tycho (which he pronounced "Teeko") was thus born into wealth and privilege, much to the benefit of the natural science of his day and ever since.

Brahe's Unusual Youth

Tycho was born near the end of 1546, the third child of the eventual twelve his mother bore. His Danish name was Tyge (pronounced "Teeg"), which he changed to the Latin form Tycho when he was in the university. Four of the Brahe children did not survive, including Tycho's twin brother. Tycho's father, Otte Brahe, was himself the younger of two brothers, both of whom rose to prominent positions of power in the kingdom of Christian III. Tycho's future, however, lay with another family.

The benefit of being "kidnapped." Unlike Otte, his brother Jorgen was childless. Jorgen waited until a younger nephew was born a year after Tycho's birth and then one day simply took Tycho home with him. Tycho was raised as the only child of Jorgen and his wife Inger. This caused no breach between Otte and his brother, surprisingly, as family ties extended more easily among nobility of the time than they do among families today.

Most of the time the sons of noblemen, once finished with formal education around the age of fifteen, acquired what more they needed to know as future leaders of Denmark by traveling to distant courts. There they learned foreign languages and new ways of thinking. They served as pages and squires to foreign lords, starting out at the bottom of the ladder as courtiers and working their way up to positions at court. Eventually they would be given administrative duties over a fief. This was the pattern that Tycho's four younger brothers followed.

Tycho took a different path, one followed by some of the men on his surrogate mother's side of the family. His new mother encouraged Tycho to continue developing his interest in academic subjects; hence, when he went abroad after his days at the University of Copenhagen, he attended foreign universities. Because he was raised in the house of his uncle, Tycho exercised an option that most likely would not have been open to him had he grown up with his biological father and mother.

Tycho studied various subjects, including astronomy and astrology, at several universities in Germany. While he was abroad the well-known incident involving Tycho's famous nose occurred. According to a much later account of the matter, Tycho had determined astrologically that an accidental incident was imminent, so he kept to his room all that day of December 29, 1566. But he came down to eat in the evening, and fell into an argument with a fellow Danish student that led to a duel. As a result of the swordplay Tycho suffered the loss of his nose. He later developed a prosthesis made of gold and silver to disguise his disfigurement.

Having followed a path different from that of most Danish noble youth, Tycho did not anticipate a life at court on his return to Denmark. Custom prohibited a nobleman from becoming a university professor, so that route was not open to him. One possibility for Tycho was to be made a canon of a Lutheran cathedral chapter, a position open to both nobles and commoners that carried a substantial income and permitted its holder to pursue a life of scholarship. Tycho was promised a canonry when he was twenty-two, although it was not actually bestowed on him until 1579, when he was thirty-three. In addition, Tycho inherited a share of his biological father's estate. He was therefore wealthy enough to follow whatever path he chose.

The empiricist emerges.   Around this time Tycho began to show how he would apply what he had learned from his teachers at the German universities. Just prior to his father's death, Tycho had been traveling abroad. He stayed primarily in Augsburg, where he revealed his concern that astronomy be based on careful and accurate observations. This conviction would later become associated with his name as an astronomer.

He did not go so far as Petrus Ramus (1515–1572), a radical thinker who came to Augsburg during the time of Tycho's visit and preached the overthrow of Aristotle. Ramus wanted to create a new astronomy based on observation and induction and in the process to eliminate all hypotheses. When he met Ramus, Tycho argued that astronomers could not dispense with all hypotheses; otherwise they would have to get rid of epicycles, uniform motion, and even the assumption that the cosmos was orderly. But he certainly agreed that new and accurate observations of the positions of the planets were indispensable to the future of astronomy.

To this end Tycho wished to improve the instruments available for identifying the location of heavenly objects. He had become convinced that existing devices were woefully inadequate. His first significant attempt to do better produced a pair of astronomical compasses that were large enough to measure reasonably small divisions of arc. Yet they were not so big and heavy that they could not be moved from one location to another.

His next try was on a grander scale, constructed for the mayor of the city. It was a giant quadrant, which was so large and heavy that it took forty men to install it on the mayor's country estate. A quadrant is a strip of metal or wood shaped into a quarter circle and mounted on a base. The quadrant strip is calibrated from 0 to 90 degrees, the sextant from 0 to 60 degrees. The degrees on both are further subdivided into 60 seconds of arc. Both are used to mark the altitude above the horizon of a heavenly object as it crosses the north/south meridian. The quadrant could be used to record the positions of stars, which do not move from night to night, or of planets and comets, which do change from one night to the next. While it was very difficult to use because of its size and weight, it represented Tycho's first step toward his goal to be able to produce measurements that were within a minute of arc of accuracy.

Marvelous Doings in the Heavens

After the death of his father, Tycho settled in for a few years on a property owned by an uncle on his mother's side—the monastery at Herrevad. This uncle, like Tycho, had spent years abroad attending universities in pursuit of an education. He had risen in the ranks of government before retiring to private life. Now he enjoyed being with others interested in matters of the intellect, especially alchemy. Tycho shared his interest in alchemy, cultivating it through experiments he conducted.

The new star of 1572.   By far the most significant event of this time occurred in the fall of 1572 when Tycho observed an object in the heavens that substantially challenged Aristotelian cosmology. Tycho wrote that after working on some alchemical experiments, he was on his way to supper when he noticed an unfamiliar star in the sky. Because he had studied the heavens since childhood, the appearance of a star never seen before, one brighter than Venus, was something that caught his eye.

Although the object became known as a new star, Tycho knew that it could not possibly be a star. Aristotle had taught that there was a fundamental difference between the heavenly and earthly realms. The heavens were eternal and changeless—stars did not suddenly appear. Nothing ever came to be or died away in the heavens. That kind of change, called generation and corruption, only occurred in the terrestrial region below the lunar orb. Therefore, anything in the heavens that appeared and then disappeared again was assumed to be located below the moon.

This view of Aristotle was still very much in vogue in Tycho's day. It had lasted so long because it was essentially correct. The only real challenge to such a view was the occasional appearance of a comet. Comets moved against the background of the stars as did planets, but to test whether the comet was beyond or below the moon was no easy task. As for the "new star" of 1572, anyone who was aware of its existence would assume that it was below the moon.

Because of its novelty Tycho gave this new star his full attention and soon determined that it did not move against the background of other stars from night to night. It did not, in other words, behave like comets or even like planets. This was a strange object indeed, especially if it was closer to us than the moon, as everyone assumed.

Although it was difficult to do with precision, there was a test he could try that might give him some information about the location of the object in comparison to the moon. The ancients knew that the heavens rotated around the Earth once each day, so two observations of the object made at different times of the night would very slightly change the relative position of the object against the background of the stars, as long as it was close to us. (The idea here can be simulated by holding a thumb in front of the face and noticing the shift in its position against a background when it is viewed first with one eye closed, then with the other. The shift in position is greater the closer the thumb is to the face.) For the moon this is called diurnal lunar parallax. It is constantly changing and even at its maximum it is around a degree for the moon. So any object closer than the moon would show an even greater parallax.

But this one did not. In fact, it showed no change whatever. It was possible that the object was moving very slowly in such a way that it just happened to cancel out a parallactic shift, so Tycho checked it over several nights. Still no change at all. That meant that it had to be farther away than the moon. When he continued to compare its position to other stars over several weeks and its position showed a complete lack of change, Tycho concluded that it was actually as far away as the stars themselves. That was a problem. Here was a star that was newly born. Apparently the heavens were not as changeless as Aristotle had thought.

The new star was bright, to be sure. To the trained eye it could even be seen during the day. But it was not something that was noticed by the casual observer. A few others in Europe had noted the appearance of the new star, although none had studied it as Tycho had. When Tycho went to Copenhagen early in 1573, no one there had heard about it. A physician on the faculty of medicine could not believe that his colleagues could have overlooked something this significant and originally concluded that Tycho was joking.

Going public as a scholar. This same professor, once convinced, urged Tycho to publish his findings. This was a major decision for Tycho because publication was not an approved activity for a person of Tycho's class and rank. He should not behave as a middle class scholar; rather, he should be pursuing the normal goals of a Danish nobleman. Tycho therefore merely wrote up a straightforward report of his findings. Then, after seeing what others were writing about the new star, he reworked his report into a book. In his book he criticized the others for describing the object as a comet and he made the revolutionary implications of his discovery unmistakably clear. He also confronted head on the issue of publishing as a nobleman. He wanted to achieve eternal glory, he said, not the passing fame brought by success in military combat. Indeed, the book made clear that he was an expert in astronomy.

These were technical matters, however. His book was not widely read and had relatively little impact. But Tycho continued to observe the new star. It began to fade by December of 1572 and continued to diminish until it finally disappeared in March of 1574. That summer Tycho moved to Copenhagen and soon was asked to give lectures on astronomy at the university. The same old problem arose—as a nobleman this was a request he could not meet. Noblemen were granted privileges, but in return they were expected to avoid intruding on the privileges of those in other stations. It was therefore not acceptable for him to give lectures at the university. So a group of noble students presented a petition to the king requesting that Tycho be allowed to lecture to them. The king added his own request to theirs, sanctioning the occasion as one requested by Tycho's peers.

Two subjects from the lectures are especially interesting. First, by basing his lectures on Copernicus as opposed to Ptolemy, it was clear that he regarded Copernicus as an astronomer with great ability who had contributed much to astronomy. But he also could not accept the Copernican system because it went against physical considerations. For one thing, it contradicted scripture, which contained several passages in which the immobility of the Earth and the mobility of the sun were clearly stated. Further, as a man skilled at detecting the parallactic shift caused by observations from two different locations, Tycho rejected the motion of the Earth because no parallax was detectable, even though the Earth radically changed its position in the course of its alleged travel around the sun. To Tycho it was only reasonable to assume that some indication of the enormous difference in the Earth's position in March from its position in September would crop up in the observations.

The other subject from the lectures was astrology. Tycho was convinced that God's vast celestial creation had a purpose beyond simple existence. That the heavens influenced events on Earth was obvious to him. However, many theologians of his day would not permit this influence to extend to the behavior of human beings. That would affect free will and thus could interfere with God's plan of salvation. For this reason they disapproved of astrology.

Once again, Tycho did not bend to official pressure. In his lectures Tycho defended astrology by taking the position that celestial influence was not the same thing as celestial determinism. People could be affected by the stars, but that did not mean that their response was determined by astral influence. By exercising free will humans could triumph over the stars and even become masters of them.

The comet of 1577.   The late-sixteenth century was unusually blessed with great astronomical events, certainly enough to reinforce the widespread astrological assumption that ominous events loomed ahead. In the late fall of 1577 Tycho was catching fish for dinner in a pond one evening at dusk when he noticed once again a bright star in the sky. Once darkness had set in, he could see that the object had a tail—it was a comet.

Given his experience with the new star of 1572, he naturally decided to see what he could determine about its position with respect to the moon. That was a much more difficult task than it had been for the new star five years earlier because the comet was near the sun and was therefore only visible for a short time after the sun had gone down. When he first saw the comet it was not visible long enough to permit the rotation of the heavens to produce much of a change in the relative position of an object against the background of the stars. A bit later, as the comet moved away from the sun in the sky, Tycho was able to attempt a measurement of diurnal parallax.

After taking various complicating factors into account, he could observe virtually no parallax. But he waited even longer so that he could check his results. His final conclusion was that, at best, the comet's parallax was not greater than fifteen minutes of arc, four times less than that of the moon. This meant that it had to be approximately four times farther away from the Earth than the moon. Tycho determined that the comet was going around the sun above the orb of the moon. Since it had suddenly appeared, here was another example of change in the celestial realm.

But this example had other implications as well, the most profound of which emerged for Tycho a decade later. It was a long time before he wrote up a Latin publication on the comet, and this work was much more detailed than the brief report he had given the king in 1578. In mid-January of 1587 he was nearing the end of his work, but the last chapter, in which he considered the writings of others on the comet, took the most time. For reasons we will examine shortly, Tycho either had not noticed or had not taken seriously a table in one of the works on the comet that he had obtained in 1579, a work by Michael Mästlin from the University of Tübingen.

Tycho must have noticed right away that Mästlin agreed with him about the supralunary position of the comet. But only in 1587 did one of the other implications in the report about the comet become clear to Tycho. Mästlin had plotted the comet's daily distances from the Earth as it traveled around the sun and found that they varied enormously. Mästlin's table had the comet at about three times the distance from the Earth to the moon at its closest point and some twenty-five times that distance before it disappeared. There was only one conclusion possible if Mästlin's table was correct: either the comet was crashing through the physically real crystalline spheres of the planets as it orbited the sun *or there were no crystalline spheres in the heavens.* For whatever reason, in 1579 Tycho did not draw this conclusion. He would come back to it in due time, after he had become a noteworthy figure in Danish society.

Tycho's Rise to Prominence

Two years prior to observing the comet, Tycho had once again been traveling in Europe. He visited old acquaintances, made new ones, and learned what others were thinking about issues in astronomy, medicine, and other topics. One of the people he met during this trip was Count Wilhelm of Hesse who, like Tycho, was an avid builder of astronomical instruments. They had much in common. Wilhelm was a man of considerable power who also made his own astronomical observations. He had found a very slight parallax for the new star of 1572, but had determined that it was so small the star must lie beyond the moon. He shared data with Tycho and Tycho helped him find an assistant to help with his work.

The fact that the two men got on so well turned out to be more than a mere curiosity. Wilhelm sent word to the Danish King Frederick that he had in his service a nobleman with outstanding abilities as an astronomer. Wilhelm recommended Tycho to the king and pointed out the advantages of supporting him in his work. The king responded by summoning Tycho once he was back in Denmark. In 1576 he offered Tycho the use of the island of Hven, with all the rents that came from the peasants who lived there. The king had decided that he wanted to capitalize on any fame that might come to Tycho, so he acted preemptively to keep him in Denmark. In so doing, the king was placing his stamp of approval on the nontraditional path Tycho had chosen to follow.

Uraniborg. Tycho set to work to build for himself a residence where he could carry out his studies of astronomy in peace and quiet, far from the political world of the court life that he so disliked. He called it Uraniborg, or fortress of Urania, after the Greek muse of astronomy. It would take some time to complete. As part of their obligation to him, the peasants of the island owed him two days of labor per week, and he made good use of it. Even so, the observatory he constructed was not completed until the fall of 1581.

Tycho filled his observatory with new instruments of all kinds. These instruments, most of which he designed himself, were constructed with an eye to providing him with the most accurate observational data possible. This was, as we have seen, a longtime goal of his. But the controversy over the new star had taught him anew how important precise data was. He determined that he would collect the most complete and precise information about the exact positions of heavenly objects possible.

To this end he either constructed or had built a series of instruments in his new observatory. They included sextants and quadrants of various sizes. The most well known, although not necessarily the most original, was Tycho's famous mural quadrant, which he mounted on a wall facing due north and south. Using this

Tycho's Mural Quadrant

instrument he could divide the calibration not only into minutes of arc (sixtieths of a degree), but into sixtieths of minutes! It was no wonder that Tycho's data became known as the most accurate available anywhere in the world.

The armillary sphere, which is a skeletal model of the celestial sphere with the Earth in the center, was especially helpful as Tycho instructed the various assistants that came to Uraniborg to learn from him. It helped them to envision the plane of rotation of the sun and the other planets around the Earth, whose equator was tilted to this plane. Tycho also had an elaborate celestial globe built, which he then improved himself into a showpiece. By 1595 Tycho claimed that he had marked on it the positions of a thousand stars.

Tycho was also interested in alchemy. He built sixteen alchemical furnaces in the basement of Uraniborg where, we must presume, he conducted experiments. Although he referred to his many findings about metals and minerals, he left no records of his alchemical work. We do know that his motives were not the making of gold; rather, he followed the Swiss alchemist Paracelsus in the quest to use alchemy to find new medicines.

The Tychonic system.   When Tycho lectured on astronomy in Copenhagen during the fall of 1574, one of his goals was not to be held hostage to the assumptions of Ptolemy's equant but at the same time to try to convert the assumptions of Copernicus to the stability of the Earth. As we have seen, something of this had long been embraced as a hope among the Wittenberg astronomers. By 1577 Tycho was familiar with an arrangement of the planets first merely mentioned by Martianus Capella in the early Middle Ages. Capella's comment was in fact noted by Copernicus in Chapter 10, Book 1, of his book. It had the inferior planets Mercury and Venus orbit the sun as the sun orbited the Earth. So when in the summer of 1580 an unusual visitor who had actually tried to sketch out such a system appeared on the doorstep of Uraniborg, Tycho must have been very interested.

The visitor was the mathematician Paul Wittich, whom Tycho had first met fleetingly in Wittenberg. Tycho soon perceived that the mathematical techniques Wittich had devised would be greatly valuable to him in his astronomical work. For his part, Wittich had obtained Copernicus's book and had sketched in its margins how he would solve the problem of eliminating the equant, which was Copernicus's great achievement, while keeping the Earth at rest, which was Copernicus's great failing.

What Wittich did in the margins of his copy of Copernicus was to start with Copernicus's system and then modify it to accommodate a stable Earth. He sketched out a system with the Earth at the center, the moon and sun orbiting the Earth, Mercury and Venus orbiting the sun, and the superior planets Mars, Jupiter, and Saturn on epicycles that circled the Earth.

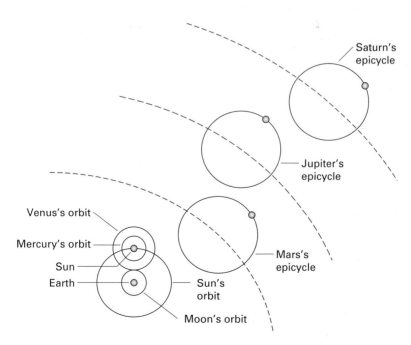

Wittich's Modification of the Copernican System

Tycho no doubt admired this achievement, but he quickly determined that it was an impossible arrangement. Tycho knew the Copernican system well. He knew that the relative distance each planet was from the sun in the Copernican system was set; for example, Saturn had to be roughly ten times as far from the sun as was the Earth. What Wittich had done, following tradition since Ptolemy, was to set the epicycles of the superior planets so that there was no wasted space. Each epicycle swung past the one above it such that it barely missed it. That gave a neatly compressed system in which the planetary spheres did not intersect, but it lost a basic feature of the Copernican system—the determination of the relative orbital distance of each planet from the sun. There had to be a better way to combine Copernicus and Ptolemy.

Sometime in the mid 1580s Tycho came up with his solution. In his new system the moon and the sun circled the Earth, and then *all* the remaining planets (not just Mercury and Venus) circled the sun. He could make this arrangement preserve the relative distances dictated in the Copernican system and still keep the Earth at rest. In fact, the new Tychonic system was observationally equivalent to the Copernican system; that is, all the lines of sight for a person on Earth were exactly the same in the two systems. This meant that, using observations alone, you could confirm both systems equally well. There was no way to distinguish one system from the other using observation alone.

There was, however, one problem. In the Copernican system the orbit of Mars around the sun was about 1.5 times that of the Earth. So when the Earth caught up

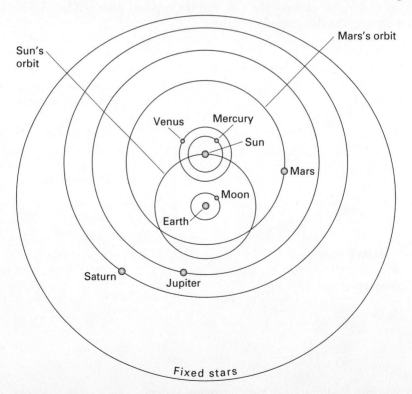

The Tychonic System

and passed Mars, it was only .5 Earth orbits from Mars. In other words, sometimes Mars was closer to the Earth than the Earth was to the sun. That caused no problem for Copernicus, but it did for Tycho. If the sun was going around the Earth, how could Mars go around the sun at the distance Copernicus required and occasionally come closer to Earth than the sun was? The only way that could happen was if Mars sometimes crossed the sun's orbital path. But that was absurd! It meant that the planetary spheres intersected.

All this was taking place during the final years in which Tycho was trying to finish his book on the comet of 1577. That was when he came across a manuscript by one Christoph Rothmann, who argued, without using measurements, that the paths of comets provided the best evidence against the existence of crystalline planetary spheres. This triggered in Tycho's mind the table in Mästlin's work on the comet of 1577. He consulted his own data on that comet and calculated its distances from the Earth, finding that they were in basic agreement with Mästlin's. He then drew the conclusion—that he later said he had been unable to entertain seriously earlier—that there were no planetary spheres.

With the problem of intersecting spheres removed, Tycho's system became viable. To accommodate the varying distances of Mars and the sun from Earth, Tycho permitted the orbits of Mars around the sun and the sun around the Earth to intersect. There were no longer any planetary spheres to worry about, so the intersection did not cause problems in that respect. And since Mars circled the sun, it could never be at the point of intersection when the sun was. That is, Mars and the sun could occupy the same position in space, but never at the same time.

Tycho had solved the basic problem that he and others had long attempted to explain: how to retain the great advantages of Copernicus's system (especially the absence of the equant) and at the same time to keep the Earth at rest. Tycho's ingenious solution, which became known as the Tychonic system, was included in the final chapter of his book on the comet of 1577, *On Very Recent Phenomena in the Aethereal Realm,* which finally appeared in 1588. Tycho considered the creation of this system to be the crowning achievement of his life's work. As we shall see in the next chapter, it was the preferred system of some Jesuit intellectuals who responded to Galileo. Further, it continued to be favored by some Dutch university scholars well into the seventeenth century.

Last years. Tycho's longtime benefactor, King Frederick II, died in 1588. His son, Christian, was only ten at the time; hence the country was ruled by four senior counselors until the boy came of age. Tycho was able to get the regents to promise that the island of Hven would revert to his heirs, but they could not bind the future king to the agreement.

Soon after the king was crowned Christian IV in August of 1596, Tycho realized that the new royal advisers did not regard him as favorably as had the regents. His most lucrative fief was recalled by the crown within a month. It became clear that the annual pension King Frederick had granted him was also in jeopardy. Tycho shut down work at Uraniborg and left the island for Copenhagen, presumably to demonstrate the drastic implications of the new policies of the crown. But it did not work. He had spent enormous sums to build what he had achieved and the crown did not regard the results as worthy of the expenditures. There was even an official request to look into a charge that Tycho had mistreated peasants on Hven. Tycho concluded that he and his family had no future in Denmark. By June of 1597 he had emigrated.

After languishing in various locations in Germany, Tycho eventually received a call in 1598 from Rudolph II, Holy Roman Emperor in Prague, to take up residence there. Rudolph appreciated scholarly achievement and he set up an estate with an annual pension, some six hours outside of Prague, where Tycho could continue his work. Arriving in the middle of 1599, Tycho began to remodel the estate so that he could do observational work. Although he did not accomplish anything here that compared to his earlier work, he was joined early in 1600 by a younger scholar named Johannes Kepler. Kepler later completed a task Tycho had undertaken for Rudolph to create new and extremely accurate astronomical tables.

Tycho died a painful death at the age of fifty-five in October of 1601. He had eaten dinner at the home of a baron, had drunk much wine, and his bladder ached. But, Kepler later related, "he had less concern for the state of his health than for etiquette" and would not excuse himself from the table. When he returned home he could not urinate and he passed subsequent days in extreme agony. Eleven days after the dinner he died.

◎ Johannes Kepler's Heliocentrism ◎

In addition to Rheticus and Copernicus himself, Johannes Kepler (1571–1630) was the other major figure who accepted heliocentrism before the sixteenth century was out. (Although it is difficult to assess Galileo Galilei's view of heliocentrism, we shall do so in the next chapter.) It is true that the heliocentrism Kepler sketched out in 1597 was augmented and modified later in his life. But this unusual astronomer never abandoned his belief in Copernicus's system.

Becoming a Copernican

Kepler came to work with Tycho Brahe in Prague in 1600. But what kind of a person was he? Where had he come from? And above all, why had he become a Copernican? In many ways Kepler's chances of becoming an astronomer were remarkably dim. His social background was the complete opposite of Tycho's and not at all likely to produce a scholar.

Early years.   Born in late December of 1571 in the southern German state of Württemberg, Johannes Kepler was the first child of a mercenary soldier and an innkeeper's daughter. The family moved around a good deal, making any kind of sustained education for their eldest son difficult, to say the least. He attended elementary school irregularly and was not in school at all between the ages of nine and eleven. Kepler does tell us that his mother took him to a high place to see the comet of 1577 and, when he was nine, his parents called him out to see an eclipse of the moon. Kepler did end up receiving an education because the duke, who had declared for the Lutheran creed, was always on the lookout during these early years of Protestantism for precocious young talent to provide future clergymen. Once the child's unusual intellectual ability was acknowledged, he was granted support to pursue an education.

His sporadic elementary schooling meant that he was behind other students of his age. But he finally completed Latin school, then attended a monastery school at age thirteen, and two years later entered a preparatory school to get him ready for the University of Tübingen. Throughout these years his introverted and unsociable personality showed itself. He became a loner, constantly quarreling with other students and unable to establish friendships. Kepler completed his bachelor's degree at Tübingen in 1588 and his master's in 1591. He then began a three-year program to prepare himself for the clergy.

At Tübingen the most important influence on him was his mathematics teacher, Michael Mästlin, whose work on the comet of 1577 turned out to be so significant for Tycho Brahe. Mästlin, himself a product of Tübingen University, had taught mathematics at Heidelberg and then had been called back to his alma mater. While he was still a master's student at Tübingen he had purchased a copy of Copernicus's book, which he read very carefully. His annotations in the work reveal that he was something of a reluctant Copernican. He praised Copernicus highly as the prince of astronomers (after Ptolemy), but he invited astronomers to complete the goal of perfecting a geocentric system so that it agreed with the phenomena as well as Copernicus's system did. That task was beyond him, he confessed, as it appeared to be for everyone else so far. Unless the common Ptolemaic hypotheses were reformed, he wrote in the margins near the beginning of the book, "I will accept the hypotheses and opinion of Copernicus."

This was the attitude that Mästlin communicated to his young student, Johannes Kepler, who became so delighted with Copernicus that he collected together all the advantages Copernicus had over Ptolemy. One of the issues Mästlin and Kepler discussed concerned a question that Copernicus left unresolved in his book; specifically, was the sun or the center of the Earth's orbit the true fixed center of the cosmos? Copernicus had centered all the planetary orbits, including the Earth's, in a point that was offset from the sun. So, which point was the center of the fixed stars? This was not a question that could be resolved at the time, but it focused the question of the place of the sun in the cosmos. To get a truly heliocentric system, the sun should be at the center of the cosmos.

Kepler wrote that he was preoccupied with three questions: why were there the number of planets there were, why were they spaced as they were, and why did they move with the speeds they did? With questions like these swimming around in his head, the young Kepler found his preparation for the ministry cut short after his second year. Over his protest, he was sent in 1595 to fill a sudden opening created by the death of a mathematics instructor of young boys in a Lutheran school located in the Austrian town of Graz. While he was there his conversion to Copernicus's system became complete. In the midst of what he called the "agonizing labor" of teaching, he stumbled on an insight that convinced him beyond any doubt that God had created the cosmos as Copernicus had described it.

The five solid theory.   The insight came as Kepler was explaining to the class why, for observers on Earth, Jupiter and Saturn came together in the sky on a periodic basis. He had drawn two orbits for the students, the orbit of Jupiter nested inside the

Inscribed and Circumscribed Triangles

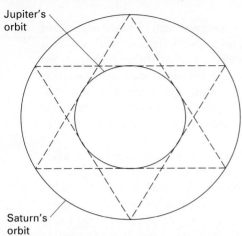

Jupiter's orbit

Saturn's orbit

orbit of Saturn, and he inserted triangles that were inscribed in Saturn's orbit at the same time they were circumscribed around Jupiter's. As he was using the properties of the triangles to explain the periodicity of the conjunctions, he was hit by a revelation. What if the Creator had set the number of planets and their distances from the center of the cosmos by nesting them using mathematical figures? He would have used perfectly formed three-dimensional figures instead of two-dimensional, and it was known already in antiquity that only five three-dimensional shapes formed perfect solids—solids whose sides were equilateral polygons. They were the tetrahedron (4 equilateral triangles), cube (6 squares), octahedron (8 equilateral triangles), dodecahedron (12 equilateral pentagons), and icosahedron (20 equilateral triangles).

If God had utilized the five regular solids, then clearly he had made six planets, because Mercury's orbit would be inscribed inside the first one and Saturn's would

Kepler's Five Solids

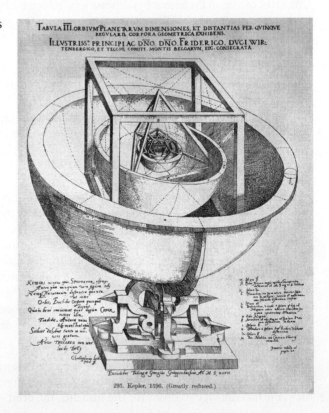

295. Kepler, 1596. (Greatly reduced.)

be circumscribed around the fifth. If there were six planets, the Earth must be one of them. The cosmos was Copernican!

Yet there were even more astonishing results to come. The shapes of the solids would determine the distances the planets were from the center, which was occupied by the sun. But in the Copernican system, unlike the Ptolemaic, the relative sizes of the planetary orbits were fixed. Was there an ordered sequence of the solids that would preserve the known relative distances of the planets from the sun? Kepler quickly found that using an octahedron, then an icosahedron, then a dodecahedron, then a tetrahedron, and finally a cube worked much too well to be accidental. "Behold," he wrote to his teacher Mästlin in 1595, "how through my effort God is being celebrated in astronomy."

The Marriage of Physics and Astronomy

Kepler wrote up his results about the planets and the five regular solids in a book entitled *The Cosmographic Mystery*, which appeared in 1597. He had answered two of the three questions that had plagued him, why a certain number of planets existed and why they were spaced as they were. Near the end of his book he turned to the third question—why the planets moved at the speeds that they did.

He did not successfully answer the third question in this work, but in taking it up he assumed that the sun somehow affected planetary motion. With this assumption he made clear that his understanding of Copernicus flew directly in the face of the interpretation Osiander had given it in his preface to *On the Revolution of the Heavenly Orbs*. Osiander had tried to assure the reader that Copernicus did not intend an actual motion of the Earth because that would insert physics into astronomy. Kepler now became the first major astronomer to demand that physical considerations in fact become part of astronomical explanations.

Kepler and Tycho. There were numerous Protestants in Graz, but Austria itself was a Roman Catholic land. The Counter-Reformation finally caught up to the village and in the summer of 1598 the Lutheran school was closed. Near the end of September in 1598 the order came for the teachers of the school to leave. Kepler, now married to a local woman, was allowed to come back because he had powerful Catholic friends and the archduke was pleased with his discoveries. But the outlook was clearly not good for the long term.

The next summer Kepler heard that Tycho Brahe was now in Prague as the imperial mathematician to Rudolph II. He knew that Tycho had spent his life gathering the most accurate data to be had anywhere. He had sent Tycho an introductory letter in 1597, which was acknowledged the following year. But then Kepler had inadvertently been drawn into a bitter dispute between Tycho and another astronomer, and now he had to placate the Danish nobleman. Letters crossed in the mail, but the end result was that Kepler made the bold decision to present himself to Tycho in Prague. This he did at the beginning of January 1600.

Although Tycho welcomed Kepler as "a highly desired participant in our observations of the heavens," Kepler found out soon enough that he was not regarded very highly. Kepler needed Tycho's data and he was not getting access to it. For his part, Tycho needed Kepler's mathematical genius to clear up certain problems in his system. By April, Kepler's dissatisfaction with his role forced a crisis in the relationship. Because of their mutual need for each other, the two were ultimately able to establish a workable, if not always amicable, relationship.

Tycho gave Kepler the toughest problem facing all astronomers—the problem of Mars's orbit. It had proven to be the most difficult planet to fit into a circular orbit around a center. But with Tycho's accurate data Kepler made some initial progress, and with surprising results. Like other planets, Mars appeared sometimes to speed up and at other times to slow down as it traveled in its path. As noted, Copernicus refused to concede that the planets experienced actual change of speed. That violated the requirement that planets experienced only perfect uniform circular motion. He had banished the equant point, explaining the speeding up and slowing down using epicycles, so that the change of velocity was only apparent.

Kepler, however, had become convinced that the sun was somehow involved in the motion of the planets. If the sun was not exactly at the center of Mars's orbit, then there would be times when Mars was closer to the sun than at others. Any involvement of the sun with Mars would then explain why it moved quicker when nearer to the sun than when far away from it. So, breaking with both the master and those who thought that Copernicus's great achievement was to rid astronomy of the equant, Kepler accepted the nonuniform motion of Mars in its circular travels around the sun.

Alone in Prague.   After Tycho died, Kepler continued to work for another ten years as imperial mathematician to Rudolph II. His primary responsibility was to produce the astronomical tables Tycho had begun. But with greater access to Tycho's data, he relentlessly pursued the Mars problem. In his mind that task had separated itself into two parts, an accurate mathematical description of Mars's orbit and a physical explanation of what caused Mars to move as it did.

Regarding the latter part, Kepler's reading of material on magnetism encouraged him to envision a magnetic force emanating out from the sun, which, because the sun was rotating, pushed the planets around in their orbits. He assumed that such a force would get smaller as the distance from the sun increased, and if the force dropped off with distance, then he inferred that the velocity of the planet would too. Since he envisioned each possible line drawn from the off-center sun to Mars as varying in this way, then all the lines taken together constituted the area of the circle.

Kepler reasoned that if the irregular motion of Mars was governed at every point by this diminishing magnetic power, then the distance and the velocity compensated each other in such a way that what stayed constant was the area swept out in equal times as the planet traveled around the sun. This was the first inkling of what later became known as the equal areas law.

Kepler had already shown that the Earth, like the other planets, speeds up and slows down at different parts of its orbit. The equal areas law fit the earth's circular orbit well, but he found that it was off by eight minutes of arc for Mars. That divergence was very small, but Kepler trusted the accuracy of Tycho's data and refused to overlook the discrepancy. He was forced to conclude something quite unprecedented: Mars's orbit was just not perfectly circular.

If Mars did not go in a circular orbit, then what shape was it? He tried an oval orbit, but, as he wrote to a Lutheran pastor in northern Germany in the summer of 1603, he did not possess the mathematical techniques to show that an oval was consistent with his conclusion about equal areas. Finally he realized that an ellipse would fit the observations and, once he had convinced himself that an elliptical orbit was consistent with his areas law, he was able, finally, to celebrate his arrival at the end of an incredible path filled with tortuous calculations. The planets go around the sun in elliptical orbits and they do so because the sun governs them. This result has become known as Kepler's First Law, while his Second Law refers to the result concerning equal areas. Kepler's was a truly heliocentric system—the sun had to be the controlling center of the planetary orbits.

Tycho's heirs were not pleased with the conclusions Kepler had come to and they blocked the publication of his results for a while. But eventually a compromise was worked out. Tycho's son-in-law inserted a preface warning readers not to be misled by the physical claims in the book that differentiated it from Tycho. The title of the book, which appeared in 1609, was *The New Astronomy: Based on Causes, or Celestial Physics.* Kepler could not have been more clear about the new manner in which he felt astronomy should be practiced. Although he had broken new ground that would define the future of astronomy, not everyone in his time followed his lead. That included Galileo Galilei, whose relationship to Kepler we shall discuss in the next chapter.

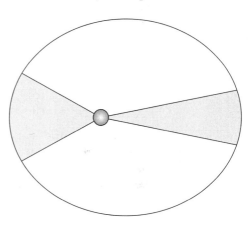

Equal Areas Law for an Ellipse

Later Years

Astronomy was not the only subject of interest to Kepler. He did fundamental work in optics, showing for the first time that the optical image is formed on the retina of the eye. Of course the emperor sought his opinion from time to time on various astrological matters. And he remained interested throughout his life in theological issues.

Up to this point in his career his choices were profoundly affected by his situation as a Protestant living in Catholic lands. That would not change soon, in spite of his departure from Prague on the death of the emperor in January of 1612. Kepler had unsuccessfully tried to obtain a position at his old university in Tübingen, but Rudolph's successor had continued his appointment as imperial mathematician and he received an invitation to pursue his work in the Austrian city of Linz.

Science and Mysticism

The NATURE
of SCIENCE

The motivation for studying natural science does not come from one single starting place; rather, it can spring from any number of sources. As is evident from the life and work of Johannes Kepler, that motivation includes the drive to find unity and harmony that has traditionally been associated with mysticism. What appears remarkable to us is that, as a result of his mystical convictions, Kepler was so successful in finding three laws of nature that have stood the test of time and can today be found in any physics or astronomy textbook. But, as Albert Einstein said, "The most beautiful experience we can have is the mysterious. It is the fundamental emotion that stands at the cradle of true art and true science. Whoever does not know it and can no longer wonder, no longer marvel, is as good as dead."

Just before the death of the emperor made it possible for him to leave Prague, his children had contracted smallpox, his wife had died of typhus, and the land was torn apart by religious turmoil. And even when he arrived in Linz, he found that he had not escaped all religious dissension. The local Lutheran pastor, heeding those in Tübingen who had prevented Kepler's appointment there because of his alleged friendliness with Calvinists, refused to give him communion.

The harmony of the world.   Not everything went badly for Kepler. He soon married again and had several more children, bringing the total to twelve. Then, when his mother was accused of witchcraft in the region where Kepler had grown up in 1617, Kepler successfully defended her at her trial.

A major part of his duties in Linz was the completion of the astronomical tables on which he had long worked. That was laborious and tedious work, and Kepler eased his mind by turning to a subject dear to his heart—cosmology. He had conceived of a work on this subject while still at Graz, but had not been able to pursue it. It was to be a summary of his cosmic vision, a natural philosophical and theological exposition of how everything fit together in God's universe.

Kepler is perhaps the best example of how far from the cautious view of some late medieval minds natural philosophers had come. Nicholas of Cusa in the fifteenth century had appealed to the properties of mathematics to show how inaccessible the infinite wisdom of God was to humans. Kepler did the opposite. For him, God was a geometer or, as he himself put it, "God is geometry itself." By delving into the intricacies of mathematical relationships, especially as the deity employed them in creating the world, humans could catch a direct glimpse of the mind of God. Doing astronomy was for Kepler a religious experience.

In *The Harmony of the World*, completed in 1618, Kepler explored the wondrous relationships of mathematics as they are found in geometry, music, astrology, and astronomy. Like the ancient Pythagoreans, he was impressed that musical harmonies, the relation between tones in the musical scale, could be expressed as simple numerical ratios of whole numbers. By his day seven such ratios were acknowledged (1:2, 2:3, 3:4, 4:5, 5:6, 3:5, and 5:8). Kepler found a rationale why these seven were the only ones possible in the properties of certain polygons, namely those that could be constructed with a compass and ruler.

Kepler then proceeded to seek the presence of these harmonies in human experience and in nature. Some of his efforts failed. For example, he thought that perhaps the periods of revolution of the various planets, or possibly the ratio of the perihelion (the closest point a planet comes to the sun) to the aphelion (the farthest point from the sun) might exhibit some of these ratios, but they did not. Finally, he discovered that the varying angular velocities of the planets as they go around the sun did fit into the musical ratios; that is, the ratio of the angular velocity at aphelion to that at perihelion for the planets fit various chords of the seven musical harmonies. And that wasn't all. If he compared pairs of planets he found even more harmonies. The spheres of the planets were creating music as they turned, evidence of "the delight of the divine Creator in his works."

The most well-known result from the book, however, had to do with Kepler's answer to the third question he had posed to himself many years before. Why do the planets go at the speeds that they do? He was now able to show that if he compared the periods of revolution of the planets to their mean distance from the sun, there was a ratio among whole numbers that captured the relationship. It was not an obvious ratio to be sure, but a man such as Kepler possessed the patience to seek until he found. He discovered that the squares of the periods of revolution of the planets were directly proportional to the cubes of their mean distances from the sun. This relationship, $T^2 \propto d^3$, was later culled from his work and dubbed Kepler's Third Law. Kepler himself did not list the results he came to as his three laws of nature.

Farewell to Austria.   The great project of the astronomical tables undertaken initially by Tycho Brahe was finally brought to completion in 1624. Included were Tycho's catalog of one thousand fixed stars; planetary, solar, and lunar tables; and other material. Just as the printing began, the Lutherans in Linz felt the heavy hand of the Counter-Reformation descend. Kepler was permitted to leave Austria, where he had spent the majority of his life, in order to have the book printed in Germany. The Rudolphine Tables were printed in Ulm in 1627, and Kepler eventually settled his family in Sagan for his remaining days.

From this time comes a work that he had begun while a student in Tübingen, and which he had completed by 1609. *The Dream,* which concerns a trip to the moon and a description of the motions of the heavens from that vantage point, is an early work of science fiction that in its own way supported a heliocentric vision of the cosmos. It was printed after Kepler's death, which occurred in November of 1630.

◎ The Status of Heliocentrism ◎

Kepler's work was highly technical and composed in a tortuous style that did not lend itself to easy comprehension. He himself suspected that few would read the glorious results he had uncovered. At the beginning of Book V of *The Harmony of the World* he wrote that his book could wait a century for a reader since God had waited six thousand years for him, a witness who understood the subtleties of Creation.

At the beginning of the seventeenth century only Kepler unambiguously and enthusiastically embraced heliocentrism. In spite of his continuing work, geocentrism remained by far the dominant viewpoint among educated people and natural philosophers alike. Not until a champion of heliocentrism emerged, an astronomer who wrote for and in the language of the people, would the issue of heliocentrism come to a head. And when it did, it became clear that dislodging the Earth-centered view of the cosmos would not be at all easy.

Suggestions for Reading

Kitty Ferguson, *Tycho and Kepler: The Unlikely Partnership That Forever Changed Our Understanding of the Heavens* (New York: Walker and Co., 2004).

Owen Gingerich, *The Book Nobody Read: Chasing the Revolutions of Nicolaus Copernicus* (New York: Walker and Co., 2004).

Victor Thoren and John Robert Christianson (contributor), *The Lord of Uraniborg: A Biography of Tycho Brahe* (Cambridge: Cambridge University Press, 1991).

James R. Voelkel, *Johannes Kepler and the New Astronomy* (New York: Oxford University Press, 2001).

CHAPTER 6

───────○───────

Galileo Galilei: Heliocentrism Gains a Champion

By the end of the sixteenth century the idea that the Earth revolved around the sun was rejected by most of the people who considered it. But that would change by the time another century had passed. Two of the numerous reasons why heliocentrism gained credibility during the seventeenth century had to do with the life and work of a fascinating man from Italy, Galileo Galilei.

First, Galileo was one of the first natural philosophers to explain the Copernican system in terms that most literate people—not just astronomers—could understand. From his annotations in his copy of Copernicus's book, it is clear that Galileo's own interest had little to do with the technical details of the work. He focused on the cosmological implications. And when he wrote about the nature of the cosmos, he rarely chose the inaccessible scholarly language of Latin; rather, he wrote in the Italian vernacular. Further, Galileo often chose to frame his discussion in the form of dialogues carried on among several interlocutors. This gave the work a broad appeal because the discussants used argument, satire, humor—any tools of persuasion they could.

Second, Galileo did not shy away from controversy; on the contrary, he seemed to revel in it. Promoting Copernicus's unusual ideas was in itself enough to bring him and his work quickly to the attention of the public. But this kind of recognition was only enhanced by another skill Galileo possessed—he was good at self-promotion. This meant that he was able to associate with leading figures of Italian public life. As a result they too became involved in the controversies surrounding heliocentrism.

Galileo's role in the history of science includes more than his promotion of the Copernican system at a time when it was not widely accepted. He made important discoveries in physics and mathematics as well. But, largely because of the clash between Galileo and the Roman Catholic Church, the world knows him best as the champion of heliocentrism.

◎ Galileo's Early Career ◎

Galileo Galilei was born in Pisa in 1564. Although the Galilei name had once been more influential in Tuscany than it was at the end of the sixteenth century, it still commanded a certain respect. The family supported itself by selling valuable cloth, a respected enough calling even if Galileo's father, Vincenzio, had had to step aside from his chosen profession as a musician to take up the business.

Galileo was the oldest of the Galilei children. When he was eight, his father left Pisa and returned to his native city of Florence. He left his son and his wife in the care of a relative until just before Galileo's eleventh birthday, when the family joined the father in Florence. Galileo received some of his early education in the monastery in Vallombrosa, even becoming a novice there. His father could not afford a better education for him, but he still had higher hopes for his firstborn child than a clerical career. If Vincenzio had his way, Galileo would attend university and become a physician.

From Pisa to Padua

The university at Pisa had been reestablished under new statutes in 1543 and since then had become the best institution of higher education the grand dukes of Tuscany had to offer. While it could not compete with the venerable Italian universities in Bologna and Padua, it nevertheless attracted a sufficient number of students and faculty to become respectable.

The family stayed at the house of a relative in Pisa so that Vincenzio could afford to send his teenage son to the university. But things did not work out as the father had planned. Although he had matriculated as a medical student, Galileo just was not interested in becoming a physician. In 1585, at the age of twenty-one, Galileo left Pisa without a degree and came home. From references he made later in his life to the behavior of customers, it is likely that for a time he helped his father in the family cloth shop.

First mathematical studies.   We do not know exactly how or when Galileo became fascinated by mathematics, but his earliest biographers tell us that it was not through courses at the university. Mathematics was a minor subject at Pisa. One suggestion is that Galileo was able to make contact with a mathematician at the Tuscan court who introduced him to the works of Euclid and Archimedes and also to the study of perspective. This teacher, Ostilio Ricci, helped Galileo persuade his father to let him continue his study of mathematics with an eye to becoming a professor.

In 1587 Galileo traveled to Rome, where he met a leading Jesuit mathematician named Christoph Gau, known as Clavius. Galileo clearly was not shy about his own abilities in mathematics; by the following year he had established a brief correspondence with Clavius about mathematical issues. He also wrote to another leading figure in the field, the Archimedes scholar Guidobaldo del Monte. He impressed Guidobaldo greatly with his mathematical knowledge, enough to win his enthusiastic help in trying to land a position somewhere as a professor of mathematics. As it turned out, that happened in 1589, when he accepted a position at his former university in Pisa.

Galileo earned a very small salary at Pisa. He had hoped for a better position, but had not yet published anything that would establish his reputation sufficiently to obtain a post at one of Italy's more prominent universities. When his father died in the summer of 1591, Galileo's desire to land a better position increased. He was now the head of the Galilei household, and he had assumed the responsibility to provide the dowry for any of his sisters who should marry. Fortunately, Guidobaldo, who had access to influential people, continued to act behind the scenes on Galileo's behalf. As a result, after only three years at Pisa, Galileo was offered a higher-paying position at the University of Padua in the fall of 1592, which he was delighted to accept.

The University of Padua had been founded in 1222 and had since established a venerable reputation in law, theology, astronomy, and especially in medicine. From the beginning of the fifteenth century to the end of the eighteenth, Padua was under the jurisdiction of the Republic of Venice. The new ideas of Renaissance humanism that filled the air were much appreciated in the Venetian republic. Padua in particular was a place that valued free and independent thinking. The republic was known for its willingness to resist the authority of Rome, so the incentive Galileo had to go to Padua was not merely financial.

Considerations of Copernicus.   As a professor of mathematics at both Pisa and Padua, Galileo had to give a basic course on astronomy. In such a course he was expected to cover how the ancients had accounted for the motions observable in the heavens. Primary among the systems he would have communicated to his students were those of Aristotle and Ptolemy. While he was aware of the more recent system of Copernicus, we know that he did not publicly endorse it in his lectures.

That is not to say that he did not discuss Copernicus at all. In a manuscript intended only for pupils, Galileo noted that there were those who said that the Earth moved, but he then reviewed the reasons given by Aristotle and Ptolemy why that could not be the case. Since this anti-Copernican stance was expected, it is not possible to determine from the manuscript what Galileo's personal view was.

There are two references from 1597, however, that seem to confirm that Galileo was not only a convinced Copernican, but that he had been so for some time. The first reference comes in a letter to a professor of philosophy at his former university in Pisa. The professor had just published a book in which he came across as unalterably opposed to Copernicus. Galileo stated that he personally thought the system of Copernicus was more probably true than the systems of Aristotle and Ptolemy. He proceeded to argue that some of the professor's objections to Copernicus were not well founded.

The second reference of 1597 is where Galileo asserted that he had long been a Copernican. Johannes Kepler had sent two copies of his *Cosmographic Mystery* to Italy to be given to those who would be able to appreciate the work. By fortunate chance one ended up in Galileo's possession and he ascertained immediately from reading the preface that its author was a Copernican. In his letter to Kepler thanking him for the work, Galileo said that he rejoiced at having found someone else who also had discovered the truth. He then added, in what many scholars regard as an exaggeration, that he had accepted the heliocentric doctrine many years ago. He had not discussed Copernicus publicly, he said, because so many ridiculed his system.

When Galileo addressed the public he did not identify with any of these alleged sympathies for Copernicus. Seven years after the letter to Kepler a supernova occurred, which was seen as a sequel to the new star of 1572 that Tycho Brahe had discussed. Galileo understood clearly that the new star was beyond the orb of the moon and he mused that in the arguments surrounding the new star, the Earth's motion around the sun might be pertinent. What he likely meant by this is that perhaps this new star might display a small parallactic shift to the careful observer; if so, such a result would be consistent with an Earth that was moving around the sun.

In a work published a year after the new star appeared, a work to which Galileo did not attach his real name, he ridiculed the treatment of the new star by an Aristotelian who did not understand the basics of astronomy. The work explained in a condescending tone the significance of parallax and it contained favorable references to Copernicus's doctrine. But when, in spite of all the attention the new star received, no small parallactic shift had been observed, Galileo thought better of casting Copernicus in a favorable light. In the second edition of this work, which appeared a few months after the first, Galileo depicted the Copernican doctrine in disparaging terms. While it is hard to be certain what Galileo believed about heliocentrism in these years, it is clear that he was willing to consider it even if he had not yet uncovered any evidence to confirm it.

Early Work on Motion

We have access to several unpublished manuscripts on motion that Galileo wrote during the Pisa and Padua years. This subject was normally the province of philosophers, so Galileo's investigations were somewhat unusual for a beginning young professor of mathematics. They are an early indication that Galileo would not feel bound by the traditional division of academic subjects.

The Leaning Tower of Pisa.   Galileo's seventeenth-century biographer, Vincenzio Viviani, originated the legend of Galileo and the Leaning Tower of Pisa. The story goes that Galileo demonstrated to observers from the university that objects of the same material but different weights, when dropped from the tower, did not, as Aristotle taught, fall at rates of speed proportional to their weights. Rather, they fell at the same speed. Unfortunately, we cannot simply accept Viviani's word on this score because at other times he has proven less than reliable. Further, there is no other record of what surely would have been quite a public spectacle.

In fact, others had long before suggested that Aristotle was wrong to assert that the speeds of falling bodies varied according to their weights. What Galileo believed was not that all objects fell at the same speed; he thought that things were more complicated than that. He thought that every material would accelerate as it fell until it reached the speed characteristic of that material. It would then continue to fall at that uniform speed without accelerating further. So if Galileo did drop objects from the Leaning Tower, it would not have been to demonstrate that all objects fell at the same rate.

Galileo's rejection of Aristotle's claim about free fall was cleverly done. Imagine, he said, two stones, one twice the weight of another. According to Aristotle the heavier one will fall twice as fast as the lighter one. So tie them together with a strong string. What would Aristotle say would happen now? He might say that the joined stones would fall slower than before because the smaller one will retard the larger one's speed. Then again, he might conclude that the joined stones will fall three times as fast as the small one by itself because the combined mass is now three times greater than that of the smaller one. Clearly, Aristotle's analysis was flawed.

More conclusions about falling bodies.   Another conclusion Galileo came to occurred after he had left Pisa for Padua. It had to do with the motion of pendulums. By 1602 Galileo had been at Padua for a decade. In that year he wrote a letter to his supporter Guidobaldo in which he tried to convince him of something that runs counter to intuition. Suspend a ball on a string, draw it to one side through an arc less than or equal to 90 degrees, and release it. Its period—the time it takes to swing back and forth—will take the same time regardless of the angle it swings through. Galileo is thus credited with noting the isochronism of the pendulum, a discovery that had important implications for measuring time intervals.

In a letter of 1604 Galileo described another new discovery regarding freely falling objects. He declared that the distance traversed in equal times increased in accordance with the sequence of odd numbers. As shown in the illustration, if an object falls for four seconds, then in the first second it traverses 1 unit of distance, in the next second 3 units, in the third second 5 units, in the fourth second 7 units, and so forth (1, 3, 5, 7, . . .). Galileo then recognized that the *total* distance traversed after a given number of seconds was given by the square of the elapsed time; that is, by the end of one second the object had fallen through 1 unit of distance (1^2), by the end of the two seconds it had traversed $1 + 3 = 4$ units (2^2), and by the end of the third second $1 + 3 + 5 = 9$ units (3^2) of distance. This meant that the distance a freely falling object traversed was proportional to the square of the time, or $d \propto t^2$.

Actual and virtual experiments.   Scholars have argued about whether the results Galileo came to were the outcome of experiments that he actually conducted or whether he relied more on so-called thought experiments. It is likely that Galileo employed both approaches and that the conclusions he came to were affected by both kinds of endeavor. Consider the following examples of Galileo's reasoning.

Beginning around 1605 Galileo tried experiments of various kinds with balls on an inclined plane. For example, he rolled a ball down an inclined plane that was set at the edge of a table, putting ink on the ball so that when it hit the floor it would

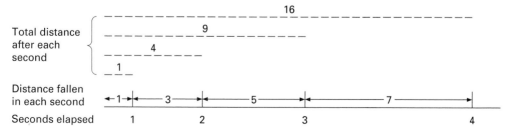

Distance-Time Relationship for Freely Falling Objects

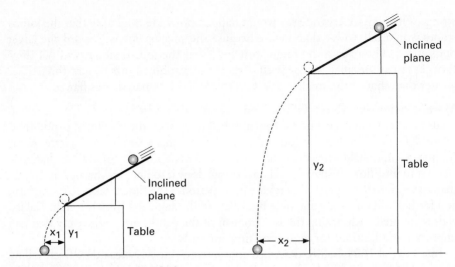

Ball Rolling Down an Inclined Plane

leave a mark. This gave him the horizontal distance the ball had moved away from the inclined plane. He also measured the vertical distance that the end of the plane was from the floor. By elevating the inclined plane apparatus to various heights above the floor, he was able to compile a set of data that correlated horizontal distances (x_1 and x_2 in the figure) and heights (y_1 and y_2).

Clearly, Galileo was depending on actual experimental results here. And from the data of these experiments Galileo was able to infer several important things. First, his measurements showed that the heights were proportional to the squares of the horizontal distance, that is, $y \propto x^2$. This correlation he recognized to be the one found in a parabola. Second, since the horizontal distance traversed during free fall was directly proportional to the time, he could substitute time for horizontal distance. That would then give $y \propto t^2$, which confirmed his earlier result about the distance fallen being proportional to the square of the time. Finally, Galileo confirmed what others had already suspected; namely, that Aristotle's followers were wrong to hold that in projectile motion the impressed force caused the object to move in a straight line until the force was expended, and that only then did the object fall straight to the ground. If the path was parabolic, then obviously two different and independent motions, one horizontal and another vertical, acted simultaneously.

But Galileo did not always actually perform the experiment on which his reasoning was based. In fact, in his thought experiments Galileo reveals another of the important insights for which he is known. By *imagining* how things would be in certain circumstances Galileo defined what is called the ideal case. That is, he imagined how something might behave in nature when there were no incidental or accidental factors present. Imagine, for example, how something might be moved if it rested on a surface and in a medium that exerted no frictional resistance at all. By doing this Galileo came to one of the most central insights in the history of science prior to that time. He concluded that under such circumstances any force, however small, would set the object into motion and that without friction to slow it down, the object would simply keep moving. He would unpack the implications of this insight later, but for now he began to understand how important it was to consider the ideal case.

Of course there is nowhere on Earth where we encounter surfaces and media that exert no frictional resistance at all. Even such slippery surfaces as ice and such nonviscous media as air exert frictional resistance. Imagining that such factors did not play a role was not something commonly done by the Aristotelian thinkers of Galileo's day. They insisted on including all of the complications they encountered in the phenomena they were trying to explain. Galileo was beginning to assume that it was instructive to establish the principles that governed the ideal case and then to explain the behavior that is actually observed by showing how the incidental factors force the object to depart from the ideal case.

Thought experiments provided other benefits than merely uncovering the ideal case. In Galileo's critique of Aristotle's position regarding the speed of freely falling objects we have already met an example of how a thought experiment could be used to analyze the merits of a received view. By merely envisioning the cases of two rocks falling separately and then tied together, Galileo uncovered results that were contradictory. Simply imagining what would happen was sufficient to show that Aristotle had been wrong. Clearly, Galileo used thought experiments as profitably as he did the actual experiments he conducted.

The relationship between nature and mathematics.   Both kinds of experiment, especially the actual ones he conducted, caused Galileo to ponder an interesting question concerning the relationship between mathematics and nature. Before discovering the parabolic shape of a projectile's path, Galileo had not assumed that the changing realm of terrestrial nature, which was filled with accidental properties, could be described using the abstract and perfectly logical relationships of mathematics. The ancients may have used geometry to describe the motions of the heavens, but the motions of things on Earth involved too many incidental characteristics to conform to the neat and trim relationships found in mathematics. In his 1602 letter to Guidobaldo, for example, Galileo noted that geometry seemed to lose its certainty when applied to things on Earth.

Now Galileo began to realize that there were contexts in which the relationships of mathematics *did* describe nature here on Earth. It was not long before he began to insist that a proper treatment of nature should include mathematical descriptions. In fact, he mentioned Aristotle's failure to do so as one of the fundamental criticisms of the ancient philosopher's explanations of terrestrial motion. Two decades later, in a work called *The Assayer,* he would identify mathematics as the language of nature.

Galileo's financial concerns.   As mentioned earlier, Galileo earned a mere pittance at Pisa, so after three years he was happy to go to Padua for triple the salary. In 1598 he was successful in obtaining a healthy salary increase at Padua, much to the annoyance of some at the university. Regardless of what others thought, Galileo never undervalued his own abilities, and he was good at marketing them.

Mathematics and Nature

The NATURE of SCIENCE

In modern natural science it has become commonplace to assume that nature can be described accurately using mathematical equations. But Galileo's discovery that the motion of falling bodies follows a simple parabola should remind us that the relationship between mathematics and nature should by no means be taken for granted. After all, why *should* the precise relationships of mathematics fit the complex physical world?

In 1960 the physicist Eugene Wigner wrote an essay entitled "The Unreasonable Effectiveness of Mathematics in the Natural Sciences," which contains many references to Galileo. One of the points Wigner wanted to make was that the enormous usefulness of mathematics in the natural sciences is something bordering on the mysterious and that there is no rational explanation for it.

Of course for men like Galileo and Kepler there could be a theological explanation. God might be responsible. If God created both the world and the relationships of mathematics, then it makes sense that the two would match up. This explanation ultimately worked for Galileo, but of course not all people are religiously disposed. The problem that Galileo uncovered is one that continues to add to the allure and mystery of modern natural science.

It was understood that no matter what salary the university paid its mathematicians, they were free to engage private pupils to supplement their income. In Galileo's case he also became involved in practical projects that carried the possibility of financial reward. He was not above pursuing patents on his inventions to ensure additional income. Further, by bringing himself attention through his inventions, he publicized his value to the republic.

Still, Galileo's financial obligations continued to grow along with his salary. One of his sisters married in 1601 and he contributed substantially to the dowry. A young woman named Marina Gamba, whom Galileo had met in Venice, moved in with him and gave birth to a daughter in 1600. Another daughter came along the next year, and a son was born in 1606. Galileo and Marina would remain together for ten years. With a growing family of his own and the possibility of more dowries on the horizon, Galileo constantly felt the need to increase his income, and this in spite of another salary increase in 1606.

◎ Court Philosopher ◎

An obvious way of alleviating the constant negotiations over salary every time Galileo's contract at the university came up for renewal was to obtain employment from a patron. Further, the steady and sure income that came with a court appointment was not the only potential benefit. He would also be much more free to carry out investigations and to write, both of which he very much wanted to pursue.

The Move to Florence

Not just any appointment would do. Galileo had, in fact, turned down an offer from the Duke of Mantua in 1604. One of the reasons he needed to find a patron of considerable power was to acquire protection should he ever need it. He was relatively safe in the Republic of Venice, where open inquiry and expression were as liberal as anywhere on the Italian peninsula. But the experiences of others suggested that even there he could become entangled in difficulties if his views were ever regarded as too extreme.

Things had been growing increasingly touchy during the preceding decades, especially concerning the control of knowledge. In 1570 the widely known natural philosopher, physician, mathematician, and astrologer Girolamo Cardano (1501–1576) managed sufficiently to antagonize officials in the Inquisition, the institution charged with the eradication of heresies, that even his willingness to stop teaching was not enough to keep him first from prison, then from house arrest. The following year the pope established the Congregation of the Index, signaling that the content of books would be coming under greater scrutiny. In some instances merely rejecting the interpretation of Aristotle's philosophy that had been crafted by Thomas Aquinas was enough to land a book on the list of prohibited writings. And then there was the example of Giordano Bruno, the Italian philosopher who was deemed a heretic and burned at the stake near the beginning of 1600. Although it was Bruno's rejection of the divinity of Christ and the notion that the Holy Spirit was the soul of the world that caused his execution, Bruno was known to defend the Copernican system, even to reject the finitude of the cosmos. Galileo would do well to be careful.

Laying the groundwork to move from Padua. Galileo preferred, of course, to go back to his native Tuscany. The roots of the ruling Medici family went back to the twelfth century, and the current grand duke, Ferdinand, was a lover of natural philosophy.

The first breakthrough came in 1605, when Galileo was invited to use his vacations from teaching at the university to tutor the Medici prince, who would one day become the grand duke of Tuscany. Ferdinand had actually appointed Galileo to his first position at Pisa, and Galileo had made his wishes known to serve as tutor to the prince as early as 1601. Ferdinand now deemed the prince old enough to become familiar with the wonders of nature and mathematics.

When Grand Duke Ferdinand died in 1609, the nineteen-year-old prince became Grand Duke Cosimo II. Galileo made inquiries through an intermediary about the possibility of a court appointment, but he simply was not yet distinguished enough to expect that they would be successful. He concluded that he would once again have to renew his contract to teach at Padua.

What changed his prospects was the arrival of news of the invention of a new instrument—the spyglass, or telescope—which Galileo learned about in that same year of 1609. The telescope had been invented the previous year in Holland and copies were beginning to turn up in various places. Someone tried to sell one in early August of 1609 to the officials in Venice, who declined. Later that month, however, they accepted a telescope Galileo had made, which he offered to them for free. Of course, the hefty increase he then received in his salary at Padua more than compensated Galileo for his trouble. His annual salary, which the Venetian Senate now bestowed for life, was more than sixteen times what it was when he started out in Pisa.

In fact, Galileo substantially improved the telescope over others that were being produced. And while he was not the only one to turn it toward the heavens as an instrument of research, he was among the first to do so. He quickly realized that the telescope made possible Earth-shaking revelations that would astound not just the House of Medici, but the whole world. With this new instrument he undertook a new project that would land him the post in Florence he so fervently craved.

The Starry Message.   The new project was the writing of a small book. It was called *Sidereus Nuncius* in Latin. Although throughout his career he usually wrote in the Italian vernacular, he chose to write this book in the universal language of the day. The discoveries were just too profound to deliver only to Italians.

The book appeared in March of 1610 and was immediately a sensation. It described three basic discoveries Galileo made through the use of a new spyglass, which, he announced in the dedication, he had invented. Galileo meant that he had invented this particular spyglass, but many have mistakenly assumed from his words that Galileo himself was the original inventor of the telescope. The three discoveries had to do with the moon, with the stars themselves, and, most astonishing of all, with four new planets.

As for the moon, Galileo reported that its surface was different from what many philosophers believed, namely, that it was smooth, uniform, and precisely spherical. It was, in fact, very uneven, like the surface of the Earth. The telescope revealed that there were mountains and deep valleys on the moon. How could he tell this? First, with the telescope he could see that the line separating the light from the dark areas of the crescent moon was very uneven, not smooth as it should be if the surface itself were smooth. Further, there were light spots visible through the telescope within the dark region, some at considerable distance from the dividing line between the light and dark regions. These bright spots were clearly the tops of mountains, which, just as on Earth, became illuminated by the sun's rays before the actual rising of the sun.

Next came the stars. The telescope did not improve their appearance—it did not bring them anywhere near as close as it did the moon. Galileo explained this by pointing out that the telescope stripped away the fringing rays that surrounded stars, making their globes appear smaller. So its magnifying power was diminished from the start. What was astonishing to him, however, was the number of new stars that became visible through the telescope—up to five hundred within one or two degrees of arc. Finally, Galileo claimed to be able to lay to rest an old debate: was the Milky Way made up of individual stars, or was it truly a nebulous patch in the heavens? Galileo declared that it was "nothing but a congeries of innumerable stars grouped together in clusters." And not only the Milky Way—Galileo explained that the telescope resolved a nebula in the constellation of Orion and another in that of Cancer.

So far the discoveries were fascinating, although they were not truly amazing. As noted, philosophers had argued about the composition of the Milky Way for some time. That meant that many had suspected there were numerous stars in the heavens that no one had ever seen. And others prior to Galileo's observations had also suspected that the moon possessed a rough surface. The discovery Galileo saved until last, however, was a real bombshell: he had seen four new *planets*. This, he said, was the most important matter of all.

Galileo recreated for the reader his own excitement at having made this find. He narrated the discovery of these new planets as an unfolding story. He gave the dates, what he had seen, what he had first thought, and then how he had figured out that these points of light were not stars, but moons revolving around Jupiter. What was not in the narrative was that soon after making this discovery, Galileo was in contact with the Tuscan secretary of state about a work he was planning to write. What did the secretary think of the idea of naming the new planets "the Cosmic stars" for the grand duke, or perhaps "the Medicean stars" for all four Medici brothers? The secretary wrote back to suggest the latter name was better—no one could miss the allusion. When the book came out, the title page proclaimed the discovery of the Medicean stars.

Galileo provided only a few hints in the book that he harbored sympathies for the Copernican system. The reader certainly did not find an outright endorsement of the Polish cleric's heliocentric arrangement of the sun and the planets. Galileo indicated his appreciation of Copernicus in passing references that might easily be overlooked. For example, he promised a sequel to his work, to be entitled *System of the World*, in which he would prove that the Earth was a wandering body, not the center of the cosmos. This meant that he believed the Earth was in motion like the other planets. In the dedication—naturally to Grand Duke Cosimo II—he observed that the duke's new planets revolved around Jupiter, and at the same time revolved with Jupiter around the sun, the center of the cosmos. And in the text itself Galileo mentioned that the existence of Jupiter's satellites removed an objection to the Copernican system, namely, that the Earth was the only body to have a moon.

In spite of these passing references the work was more sensational than it was truly revolutionary. Nothing he had seen required anything more than an updating of the inventory of the cosmos, not of its arrangement. The reader might have to conclude that the moon had a rough surface, that there were more stars in the heavens than once thought, and that one of the planets had moons. But none of that challenged the Ptolemaic arrangement of the cosmos. It just provided updated information. The only real challenge to Ptolemy was Galileo's promise to prove that the Earth moved. The other allusions—to there being two centers of motion in the cosmos and to Jupiter's revolving around the sun—were both matched in the system of Tycho Brahe. All of this would be argued about in due time. In the spring of 1610 the dust was still swirling.

The call to Florence.   Galileo's book came out in March. He took immediate steps to implement his plan to convince the grand duke to bring him home to Tuscany. In January he had cast a birth horoscope for the grand duke. He promised to send the grand duke himself the very telescope he had used to make his discoveries. He sent a copy of his book to the Tuscan ambassador in Prague so that it would come to Kepler's attention. And then in May of 1610 he wrote again to the Tuscan secretary of state to discuss his possible appointment to the Tuscan court. By that time he was able to include the hearty endorsement Johannes Kepler had sent to Galileo. Kepler was happy to sing the praises of Galileo's work and had written back in less than a week to say so.

Galileo did not seek an increase in salary, but he did request that his appointment bestow on him the title of philosopher as well as mathematician. In fact many of the topics he addressed in *The Starry Message* concerning the existence and constitution of heavenly bodies were considered the philosopher's subject matter. As a mathematician he was supposed to confine himself to the description of the heavens, so he was technically poaching on the academic turf of others when he gave arguments for the existence of new stars, the Medicean planets, or mountains on the moon. And since philosophers occupied a higher rung on the social scale, an appointment as a court philosopher would instantly raise his status.

Of course Galileo made clear what he would do as the grand duke's philosopher. *The Starry Message* was just the beginning. He outlined the works he would be able to publish once he enjoyed the free time for research that freedom from regular teaching duties would permit. And all of this would redound to the glory of the grand duke. His petition worked. In June the secretary wrote to tell him that he had the position and, amid considerable resentment on the part of those who had just raised his salary in Padua, Galileo resigned his university post. The appointment was formally made in July and he moved to Florence in September.

Early Triumphs of the Ducal Philosopher

Life in Florence, even family life, would be very different. Although Galileo took his two daughters with him to Florence, Marina Gamba remained behind with their four-year-old son, Vincenzio. His son joined Galileo a few years later, but the separation from Marina proved to be permanent. She married another man in 1613.

The greatest difference, however, was that Galileo no longer had to teach. He continued to give limited private instruction, but that burden had also been largely lifted. He had promised the grand duke that he would make more discoveries. Before the end of the year he had made a very significant one.

A new discovery.   Galileo continued to observe the heavens with the excellent telescopes he had constructed. In December of 1610 he observed that the planet Venus had phases. This sounds like the kind of observation he had made in *The Starry Message*—new information for updating the cosmic inventory. It was not. Venus could *not* have phases under the old Ptolemaic system. This discovery provided a substantial discrediting of Ptolemy's system because the only way Venus could display phases was if it were sometimes between the Earth and the sun and sometimes on the other side of the sun from the Earth. But that would mean that Venus circled the sun.

Both Galileo and the Jesuit astronomer Benedetto Castelli—who had observed Venus's phases at about the same time Galileo did—concluded that the observed phases of Venus vindicated the system of Copernicus. As can be seen in the diagram, the phases Galileo and Castelli observed are consistent with the Copernican arrangement of the planets in their trip around the sun. At position E Venus would

The Phases of Venus

be at new phase because the portion of Venus illuminated by the sun is opposite the side seen from Earth. It would then grow to crescent at A, then to the almost full "gibbous" phase at B. It would not be visible at Q because the sun itself blocks the view from earth, but would be waning gibbous at C. At D it would again be crescent shaped, to return to new phase at E. Further, it should be a larger crescent image than gibbous image because Venus was closer to Earth when at crescent phase than at gibbous. These were exactly the phases that Galileo and Castelli observed.

There is, of course, one flaw—a major flaw—in the inference that the phases of Venus proved the Copernican system. It is that the phases Galileo observed through the telescope were equally well accounted for by the Tychonic system, in which Venus also circles the sun. As we saw in Chapter 5, the systems of Tycho and Copernicus are observationally equivalent, so it should be no surprise at all that the telescopic observations of Venus's phases served as evidence for both the Tychonic and Copernican systems.

As historian Mario Biagioli points out, the Jesuits began to lean toward Tycho's system soon after Galileo's discoveries. Galileo, however, regarded Tycho's Earth-centered cosmos as looking backward, not forward. Tycho's geocentric system was a thorn in Galileo's side; it meant that he really could not absolutely prove that the Earth was in motion. Still, after he discovered the phases of Venus, Galileo became much more open than he had been about expressing his opinions on Copernicus. His inference that the phases of Venus meant Copernicus was correct would cause difficulties for him in the developments of the coming years.

Triumph in Rome.   Galileo's fame was increasing. As a representative of the Grand Duke Cosimo II he could not be simply dismissed, although numerous people resented his presumption to tread on ground where he did not belong. From the start there were those in Padua who downplayed his achievement. With great public success came the prospect of public enemies.

For the moment, however, Galileo was on top of the world. In March of 1611 the Tuscan ambassador in Rome invited him for a visit in order to draw attention to the grand duke's recent appointment. Of course Galileo visited Clavius, the first mathematician he had contacted when he was just starting out. He was reassured when Clavius mocked a recent attack by Francesco Sizi, author of *Dianoia Astronomica*, who declared that the Medicean planets did not exist because they could not be seen by the naked eye. He was received by cardinals and even had an audience with Pope Paul V. He did not return to Florence until June, when he settled down to the work of establishing his credentials as a philosopher.

◎ The Path to Conflict ◎

Few episodes from the history of science are of greater interest than the clash between Galileo and the Catholic Church over the merits of the system of Copernicus. The early successes that brought him public acclaim also earned him enemies who did not wish to see his celebrity continue to rise. A setback to Galileo in 1616, followed by a turn of fortune for the better in 1623, all culminated in 1632 with the publication of his famous book on the two chief world systems. This work precipitated the judgment against him that his adversaries had long hoped for.

Storm Clouds Looming

Even in Tuscany, some Aristotelian philosophers set themselves against Galileo relatively quickly. The first to publish against Galileo was Martinus Horky, who tried to discredit the reliability of the telescope. We have already mentioned Sizi's attack on the existence of the Medicean stars. These attacks focused on the observations of *The Starry Messsage.*

Not long after he had his meeting with the pope, Cardinal Robert Bellarmine, chief advisor of the Holy See, sought the view of Jesuit scholars about Galileo's observations. He was told that the telescope had definitely uncovered the presence of new stars in the heavens, including the four satellites of Jupiter; further, Venus was observed to have phases. Some doubted that Galileo had demonstrated all he said about the surface of the moon, but there was little doubt that in general the observations were real. Most Jesuit mathematicians approved of them because they were pleased with the way Galileo was raising the status of mathematicians to rival that of philosophers.

The problem of Copernicanism.   Jesuit mathematicians may have endorsed the telescope, but that did not mean that they approved of Galileo's implicit endorsement of Copernicus. This was because, as already pointed out, the observations could also be seen as evidence of the geocentric system of Tycho Brahe. There was nothing about the telescopic observations that permitted one to exclude Tycho's system in favor of Copernicus's. Approving the reliability of the observations and approving Copernicus were two entirely different matters.

Things began to change as Galileo's Copernicanism became the focus. The Florentine philosopher Lodovico delle Colombe penned a manuscript entitled "Against the Motion of the Earth." Delle Colombe's declaration that the doctrine of Copernicus was contrary to the Scriptures marked the beginning of a growing concern.

This was in 1611. Near the end of 1612 a Dominican priest, who did not even know Copernicus's correct name (he called him "Ipernicus"), mentioned to Galileo that heliocentrism did seem contrary to Scriptures. Galileo most likely realized he was going to have to deal with this question. He had already crossed over into the territory of philosophers. Now he was about to take on the theologians.

The letter to the grand duchess.   One year later, in December of 1613, the mother of the grand duke raised the question anew with Galileo's friend, Benedetto Castelli, now a professor in Pisa. Castelli wrote to Galileo to tell him about the grand duchess's interest and Galileo immediately replied with his thoughts on the matter. He made it clear that this was a case when the words of the Bible could not be taken literally. In the Old Testament, for example, to prolong daylight for his army, Joshua commanded the sun to stand still (which suggested the sun, not the Earth, was moving). Galileo explained that the writer here was not taking a stand on whether it was the sun or the Earth that was moving; rather, he was portraying events in language his readers could understand. Because the matter was such a central concern, Galileo told Castelli that he could share the contents of the letter. Naturally, the letter was copied and began to circulate, sometimes among friends, sometimes enemies.

It took a year for hostile theologians—who resented Galileo's presuming to tell them how to interpret Scripture—to join forces against the grand duke's philosopher. The opening salvo came from a Dominican priest named Tommaso Caccini, who preached a sermon in Florence at the end of 1614 on the book of Joshua in which he denounced Galileo and Copernicanism as contrary to Scripture. Another Dominican, the same one who had mistaken Copernicus's name a year earlier, obtained a copy of Galileo's letter to Castelli. Early in 1615 he sent his copy of the letter to the Inquisition. Soon thereafter Caccini went to Rome himself to speak to members of the Inquisition.

When Galileo got wind of all this he became concerned that the copy of the letter being used in Rome was inaccurate, so he sent a true version to friends in Rome who would make sure it got to the right people. Against the advice of theologian friends who urged Galileo not to become involved in theological arguments about Copernicus, Galileo expanded his letter to Castelli into a treatise entitled "Letter to Madame Christina of Lorraine, Grand Duchess of Tuscany, Concerning the Use of Biblical Quotations in Matters of Science." Although not published until much later, the work represented the views Galileo held on theology and science.

In the treatise he repeated his claim that while the Bible cannot contain untruth, it must not always be taken literally. Were we to do so we would have to assign God feet, hands, and eyes, and even acknowledge that he occasionally forgets things. When the writers of the Bible used such images they were accommodating the understanding of the average person.

Galileo even included a technical argument that demonstrated why we should not take the passage about Joshua and the sun literally. To prolong the day Joshua would have to have commanded the outermost sphere of the heavens, which drove everything around the Earth once a day, to stop moving. The sun's proper motion goes against the grain of this daily motion, although very slowly. Had Joshua commanded the *sun's* motion to cease without at the same time stopping the twenty-four hour motion of the outermost sphere, the day would actually have been slightly shorter! So clearly the writer did not want us to invest a literal meaning in the words "Sun, stand thou still." The effect of this clever argument, however, might well have been simply to provoke the ire of those who did not know astronomy.

The 1616 visit to Rome. Late in 1615 Galileo became concerned that things were moving against him in Rome, so he decided to go there himself to make his own case. He had been in correspondence with people there, including the chief advisor, Cardinal Bellarmine. He and Bellarmine agreed on some things and disagreed on others. They agreed that the Bible could not err and that true science and true religion could never conflict. Where they appeared to disagree was over the status of hypotheses. Bellarmine regarded hypotheses as assumptions of convenience, in Copernicus's case, the convenience of making calculations. He did not see them as assumptions that were likely someday to be proven true. He even made his own hypothesis about the most convenient way to handle Copernicus. Say the motion of the Earth *was* clearly demonstrated. In that case we would have to say not that the Bible was wrong, but that we did not understand it. He felt himself to be on very safe ground: the Earth's motion had not been (and would not be) clearly demonstrated.

Galileo undoubtedly saw the Copernican hypothesis as something that could be—and in his own mind had been—proven true. He had a favorite argument to demonstrate the Earth's motion, which he had first thought of while in Padua. The existence of tides, he said, resulted from the contrary motions the Earth experienced between its motion around the sun and the motion on its axis. Sometimes these two motions reinforced each other and sometimes they went against each other. The result was the sloshing of the seas against the shores. So the fact that there were tides proved the Earth was moving.

Galileo was aware that this explanation entailed just one high tide and one low tide a day, whereas two were observed to occur. But he said that other factors were also involved that were responsible for the extra tidal motions. Of course, if he were right, then Bellarmine was wrong in assuming that the hypothesis of the Earth's motion would not be proven true. Confident that he could persuade the right people, he went to Rome to vindicate himself.

Things went well for him personally. He soon established that he himself was in no danger, so he decided to push for an acceptance of Copernicus. He persuaded a friend among the cardinals to talk to the pope. The pope's response was to request that the Inquisition make a formal evaluation. Caccini had identified two propositions to the Inquisition that he deemed heretical: that the sun was at the center of the world and immobile, and that the Earth was not the center and moved with a double motion. The Inquisition determined that the first proposition was heretical because it contradicted Scripture. The second was theologically erroneous.

The pope instructed Bellarmine to communicate to Galileo the judgment that he was not to teach or defend or consider the Copernican doctrine. Bellarmine met with Galileo and apparently told him that he was neither to hold nor defend Copernicus's view. At least that is what is contained in a copy of the letter Bellarmine later sent to Galileo confirming the order. Given their discussion about hypotheses, Galileo assumed he could at least continue to use the Copernican system hypothetically in his future work in astronomy. Exactly what Bellarmine enjoined Galileo to do became the center of contention in Galileo's later trial, as we will see. The point, however, was to keep Galileo quiet so that he would not raise further difficulties.

Galileo's Famous Book

Between 1616 and 1623, when a new pope was elected, Galileo publicly acknowledged that the Copernican system was not the true system of the heavens. Of course he added that neither was the Ptolemaic system, because that had been shown to be wrong. As always, he dismissed Tycho Brahe's system. That left him able to voice the fervent hope that someday the true system would be found.

These sentiments were expressed in an exchange about comets, which Galileo became entangled in during these years. Unfortunately, as the controversy unfolded he needlessly antagonized a Jesuit mathematician in Rome named Orazio Grassi. Galileo's unnecessary attack elicited a reply in which Grassi, an adherent of Tycho's system, raised anew the issue of Copernicanism. It was not that the debate about comets entailed the question whether the cosmos was heliocentric or not, but Grassi was in a real sense baiting Galileo. He knew that Galileo could not defend his Copernican views publicly.

Another visit to Rome.   The cardinal elected to the papacy in 1623 gave Galileo renewed hope because he was a friend, Mafeo Barberini. Barberini had sided with Galileo earlier in a controversy over floating bodies. He also appreciated Galileo's devotion to astronomy, although prior to the 1616 visit to Rome he had urged Galileo not to become involved in theological debates about Copernicanism. Now he was the head of the Roman Catholic Church. Naturally Galileo's hopes were raised. Perhaps he could persuade his old acquaintance, now Pope Urban VIII, to remove the prohibition against Copernicus.

Galileo visited the pope in Rome in the spring of 1624 and had six conversations with him. They remained on friendly terms, but Galileo did not get what he wanted—permission to ignore the injunction of 1616. They understood each other well. Each had a pet argument of which the other was aware. Galileo felt that the tides proved the motion of the Earth. Urban VIII believed that no matter how things seemed to be, God could have made them that way in any number of ways. The existence of tides did not mean that God had produced them by the motions of the Earth. God could have easily caused them on a stationary Earth by some other means.

Because of their discussions, and because the pope acknowledged that the church had not condemned nor would it condemn heliocentrism as heretical, Galileo felt that he could discuss Copernicanism publicly as long as he did not defend it or appear to hold it to be true. So he set out to do just that, to write a work comparing Copernicus to Ptolemy in which he discussed the merits and demerits of each *without openly endorsing Copernicus.* Once again there would be no mention of Tycho's system, even though it was consistent with all of the telescopic observations.

Around this time Galileo began exchanging letters with his eldest daughter, Virginia, a nun who had taken the name Sister Maria Celeste to honor her father's fascination with the heavens. She was cloistered in the convent at San Matteo in Arcetri, outside the city walls on the southern hills of Florence. Through her obvious devotion and concern for his well-being, and by simply being there as a correspondent, she imparted strength to Galileo to help him cope with his struggle to be true to himself and his church.

Galileo's nephew, however, caused him nothing but grief. After Galileo arranged support for him to study music in Rome, the nephew turned out to be pure trouble.

He was antireligious and much too open about it. Fortunately for Galileo, he left Rome before any serious breach occurred; otherwise, he might have dragged his uncle into an unseemly situation at a time when he could ill afford it.

Galileo's *Dialogues of the Two Chief Systems of the World.*   Although Galileo's eyesight began to fail, he finished the work comparing Ptolemy and Copernicus by late 1629. In the spring of 1630 he went to Rome to arrange for publication. While there he had a friendly meeting with the pope. Urban VIII was not pleased that Galileo had included his old argument about the tides in his forthcoming book, although he did not demand that Galileo remove it. During the process of obtaining permission to publish, Galileo learned that the pope insisted that he portray the Copernican hypothesis merely as a system capable of accounting for what one saw in the heavens. He was not to portray it as the truth.

Galileo attempted to satisfy the concerns he knew were present by casting his work as a discussion among three speakers, two of whom were named for close friends. They were the participants versed in Copernican theory who, without declaring it to be true, were capable of answering objections to it. In this manner Galileo felt that he was meeting the requirement from 1616 that he neither hold nor defend Copernicanism as true. Galileo named the third speaker Simplicio after a Roman commentator on Aristotle. He was the defender of Ptolemy and came across as possessing less ability than the other two interlocutors. With this name Galileo in fact suggested that the character was something of a simpleton.

The book rehearsed systematically and at length all of the issues pertinent to the structure of the cosmos that Galileo had dealt with over his career. He arranged the book into four days of discussion. First he developed all of the arguments against the old Aristotelian idea that the heavens were incorruptible, that the terrestrial and celestial realms were qualitatively different places.

In the second day he took up objections raised against the alleged motion of the Earth; namely, that a rock dropped from a tower on a rotating Earth would not fall directly to the base of the tower (as it is observed to do), but, in the time it took to fall, would land "many hundreds of yards east." Galileo countered this conclusion by arguing that a rock dropped from a ship that is known to be moving also is observed to fall at the base of the mast, not somewhere behind the moving mast. Galileo devoted this section of his work to criticizing Aristotle's claim that all motion requires a mover. There is one kind of motion that does not require a mover. Motion at a uniform speed in a perfect circle will continue on forever unless it is interrupted by the action of an external force.

Galileo here was creating a new physics of motion. He was maintaining that uniform circular motion was a state of being, like rest; indeed, they were equivalent states of being. Just as we do not take it upon ourselves to explain why things remain at rest, so we should not try to explain uniform circular motion. Both are natural states—they need no further explanation. Because they are equivalent states of being, we cannot distinguish when we are in one or the other unless we can compare our frame of reference to the other. They feel exactly the same. Locked in a room, we cannot tell if we are moving or not, but once we have access to the heavens we know that either we, or the heavens, are moving. That is why we do not feel the rotation of the Earth on its axis, or the revolutionary motion of the Earth around the sun—these states of being are exactly like being at rest. In developing his argument about what is called inertial motion Galileo was answering a set of major objections to the Copernican doctrine. He was explaining how it was possible that the Earth could be moving without us feeling it.

Galileo's new understanding of motion appeared to his critics to be unnecessarily complicated. They thought that the reason why an arrow shot straight up comes down close to where it was released is because the Earth beneath it is not moving, not because it shares the Earth's motion. And even in Galileo's example of the ship, where something *is* moving, the speed is not great enough to make a difference. That is why the rock falls to the base of a moving ship.

In the third day Galileo continued to give reasons why a moving Earth made sense, and then, in the fourth, he moved to the heart of his argument: the tides. They proved, he felt, that a moving Earth made sense. The Ptolemaic cosmos had already been disproved by the phases of Venus. If he was right about the tides, then Galileo here had a means of disproving Tycho as well. Of course he was not so bold as to say that he had definitively proven the motion of the Earth—that would have disobeyed the 1616 injunction. He said merely that this tidal motion followed naturally from the supposition that the Earth moved with the double motion Copernicus had assigned it. He did, however, reproach Kepler—who had associated tidal motion with the influence of the moon—for appealing to occult causes.

It is obvious why Galileo thought the tides were so central to his work. Without them he thought he had no proof that the Earth moved. He had wanted to title his book *On the Ebb and Flow of the Sea,* but the pope, who also realized what was at stake, prohibited that. The pope was not on familiar territory—he could not show why Galileo's argument was wrong. (Others, incidentally, did show much later that Galileo's explanation of the tides was erroneous.) Urban VIII allowed Galileo to include the tides because of his own conviction that God could have easily caused them on a stationary Earth. For the pope, Galileo's argument about the tides did not definitely prove the Earth's motion.

Galileo placed the pope's favorite argument at the very end of the work. There it is given to the character of Simplicio, who declared that he learned it "from a most eminent and very learned person." Galileo apparently thought that he had satisfied the pope's concerns by ending his book with an argument "before which," Simplicio proclaims, "one must fall silent." But after spending an entire book removing arguments against Copernicus, this sop to the pope's position, put into the mouth of the simpleton, revealed that Galileo did not think much of the argument at all. The pope certainly interpreted it this way and was personally insulted. Friend or no friend, Galileo had crossed the line.

◎ Last Years ◎

Galileo's final years were fraught with tribulation. His poor health did not prevent him from being ordered to Rome to stand trial, even though his physician recommended that he not travel. Telescopic observations of the sun eventually took their toll—he could no longer see well. The ordeal of the trial itself did not go well for him and it was followed very quickly by the death of his beloved daughter Maria Celeste. Were it not for his completion of work begun long before, these years would have been bleak indeed.

The Trial

Many people, including the pope, had been aware that Galileo was working on a book that dealt with Copernicus. Galileo had the sense that as long as he obeyed the letter of the injunction of 1616 not to hold or defend the Copernican doctrine, there would be no objection to his work. That is why he had not sought permission to write the book.

But once it was written, he did need to have it licensed and obtain the required permissions to print it. Licensing was up to the pope's chief theologian, who insisted that the opening and closing of the work reinforce the hypothetical nature of the discussion. Further, he indicated that the pope did not want the work to focus on the tides; rather, it should show that, as long as one set aside the truth of revelation, the mathematical hypothesis of Copernicus could make sense of the observed motions in the heavens. Galileo got approval to use censors and printers in Florence, provided they abided by these requirements. The book appeared in February of 1632.

By August publication was suspended. The pope was outraged by the way Galileo had written the book. And he severely reprimanded his chief theologian, who had licensed the work. He turned the matter over to the Inquisition with the clear understanding that Galileo would have to answer for his misdeed. The Inquisition summoned Galileo to Rome even though, at age 70, he was not well. He arrived in February of 1633 and experienced the first interrogation in mid-April.

The trial turned on the issue of whether Galileo had abided by the injunction of 1616. Galileo understood that injunction to mean that he was not to hold or defend the Copernican hypothesis and he produced the letter Cardinal Bellarmine had sent him following the 1616 meeting in support of his understanding. The inquisitors cited passages from the *Dialogues* in which, they said, Galileo had treated Copernicanism not as a hypothesis, but as truth. They cited the *Letter to the Grand Duchess Christina* as further evidence that Galileo was a Copernican.

In addition, the inquisitors produced another document from the files of the 1616 meeting that went much farther than the one Bellarmine had given him then. It said he was not to hold or defend or *teach the doctrine "in any way whatever."* Galileo indicated that he had no recollection of such a document, which was not signed by anyone, certainly not by Galileo to show that he had received it. He might have thought that those in the church who had been aware of his wish to write a work on Copernicus should have advised him against it in light of this document. All that was to no avail. The Inquisition had found legal grounds on which to convict Galileo. In his defense, Galileo continued for a while to deny that he had held the Copernican doctrine. He was shown the instruments of torture that were used on those who did not tell the truth. Eventually Galileo decided to capitulate and concede that he had maintained a belief in heliocentrism and that he now recanted such belief. The

Inquisition condemned him for teaching Copernicanism as probable even though it went against Scripture. A year later Galileo was placed under house arrest in Arcetri, near the convent where his daughter lived. Although she died of dysentery less than four months after he arrived in Arcetri, Galileo himself lived until 1642.

Final Work and Death

Once he recovered from the experiences of the trial and then his daughter's early death, Galileo was able to return to the subjects that had fascinated him in his early career. He decided to write up the work he had done at the beginning of the century on the motion of objects here on Earth. A new book, entitled *Discourses on Two New Sciences,* appeared in 1638 even though Galileo had been prohibited from publishing further. Galileo said that he gave a copy of his papers on motion to a former pupil who had visited him. He had not wanted them to perish. That pupil, who was now an ambassador in Rome from France, saw to it that they were published in Holland.

While not as sensational as his *Dialogues,* the book summarized important work that has defined his contribution to physics. It contained a treatment of the material discussed earlier in this chapter about Galileo's early study of motion, including the work on falling bodies and projectiles, the pendulum, circular inertial motion, and even his thoughts on the relationship of mathematics and nature. While other topics were also included, here was Galileo making sure that posterity would appreciate the particulars of his work on motion. Galileo described his experiments with inclined planes, by means of which he slowed down the accelerated motion of falling bodies so that he could verify his results. Needless to say, Copernicanism was not mentioned.

Galileo died in January of 1642. The last decision to be made about this great natural philosopher was how his passing would be marked. He was a celebrated figure, to be sure. And yet he was serving a sentence for a serious crime. Naturally, the Grand Duke of Tuscany wanted to commemorate Galileo with a memorable tomb befitting of a great man. Urban VIII, however, would not hear of that. Galileo was buried privately in the church of Santa Croce in Florence. A century later his remains were moved into an elaborate tomb in the same church, where his life and work have been celebrated since.

Suggestions for Reading

Mario Biagioli, *Galileo Courtier: The Practice of Science in the Culture of Absolutism* (Chicago: University of Chicago Press, 1994).

Michael Sharratt, *Galileo: Decisive Innovator* (Cambridge: Cambridge University Press, 1996).

William R. Shea and Mariano Artigas, *Galileo in Rome: The Rise and Fall of a Troublesome Genius* (Oxford: Oxford University Press, 2003).

Dava Sobel, *Galileo's Daughter: A Historical Memoir of Science, Faith, and Love* (New York: Penguin Books, 2000).

CHAPTER 7

---◎---

Natural Philosophy Transformed

During the same years that Galileo was embarking on a collision course with the Roman Catholic Church, others outside Italy developed their own alternatives to traditional natural philosophy. In this chapter we shall examine the variety of approaches that existed in the first half of the seventeenth century. By emphasizing empirical observation, experiment, and reasoning from the particular to the general (induction), and by replacing the older Aristotelian understanding of nature as organism with nature as mechanism, seventeenth-century thinkers transformed natural philosophy in several ways.

◎ British Conceptions ◎

In the first half of the seventeenth century, many in Northern Europe took up their pens in defense of new conceptions of the world in which they lived. In natural philosophy British thinkers were especially innovative in the manner in which they reacted to past approaches to understanding the natural world.

Harvey and Scholasticism

Scholasticism, the philosophical and theological systems of the Middle Ages, continued to thrive in the university curriculum, where a mastery of Aristotle's thought was still required. Indeed, fundamental work in natural philosophy leading to new knowledge was carried out in Britain, as on the Continent, from within the confines of Scholastic thought. For example, the Scholastic physician William Harvey (1578–1657), in a work of 1628 entitled *On the Motion of the Heart,* convincingly demonstrated through animal dissections that more blood passed out of the heart in an hour than the body held, suggesting that it must in some manner be reused.

Although he argued that blood moved away from the heart through the arteries and back through veins, Harvey could not demonstrate how the transfer from arteries to veins took place. But he demonstrated, again through experiment, that a transfer must occur. He tied a band tightly around the forearm, noting that the veins continued to look normal as the arteries swelled with blood. He then loosened the band to permit the flow of arterial blood down the arm, but not enough to permit venous blood to flow past the band. Now the veins became gorged with blood, suggesting that the arterial blood that had been let through had somehow gotten into the veins.

Through this and many other experiments, Harvey helped to overthrow the older view of Galen that the blood ebbed and flowed in the veins and arteries. But while Harvey's work provides an example of the successful continuation of Scholastic thought, many new currents of natural philosophy soon appeared that challenged Aristotle's worldview.

The Meaning of the Break with Aristotelianism

Almost all of the new visions of natural philosophy that made their appearance in the early decades of the seventeenth century showed themselves first through their dissatisfaction with Aristotelian Scholasticism. Galileo was therefore not the only one to express discontent with the dominance Aristotle's system still enjoyed in the university curriculum.

The difference in goals. While there were various degrees of dissatisfaction with Aristotelian natural philosophy, a new English natural philosophy of the seventeenth century represented the most profound break. It set out to accomplish something fundamentally different from what the Scholastics were trying to achieve.

What was the difference between the goals of Aristotelian thought and those of the new English natural philosophy? As historian of science Peter Dear has explained, the intent of an Aristotelian philosopher was to understand phenomena that were *already known*. The emphasis was on logical reasoning from generalized starting points that were assumed to be true. Through their generality these premises summarized what was universally known, while the process of reasoning brought out conclusions that were implicit in the premises. Awareness of these implications supplied clarity and a fuller understanding. Gaining knowledge, then, was filling out what we already know.

The growing number of critics of Aristotle in seventeenth-century England did not set out, as Aristotle did, to unearth the implications of an *assumed* general premise; rather, they claimed that it was possible and necessary to *construct* universal statements through experiments. They believed that the major premise of a logical deduction was not self-justified, as Aristotle thought. In their view Aristotle assumed such general statements because they had been reinforced over and over again in his experience of individual events. He believed, for example, that all motion required a mover because he had seen so many things moved by movers.

As a result of this very different perspective, these natural philosophers had an approach to the observation of the natural world that was dissimilar to Aristotle's. They set out to *discover* new knowledge, not to further understand existing knowledge. Their attitude was that we must not presume we already know; on the contrary, we do not yet know and we must find out. An attitude like this meant that they were not content, as Aristotelians were, merely to observe nature's ordinary course passively. Because they assumed responsibility for discovering new truth, they were aggressive. They wished to force nature to reveal secrets by creating circumstances that did not occur naturally and seeing how nature responded to them. Nature was something to be interrogated and controlled.

The impact of craftsmen and magicians.   The new attitude of English natural philosophers did not come out of nowhere. It had precedents in the sixteenth century. In Chapter 3 we considered the stance taken in medicine by Paracelsus, who in the first half of the sixteenth century enjoyed opposing the authority of established Aristotelian natural philosophers of his day. Paracelsus was perceived as a radical, in part because he refused to abide by the established social distinctions of his day that prescribed how a physician should behave. His view was similar to that of the crafts-man, who occupied a lower social position than that of university professors.

Like a craftsman, Paracelsus believed that to understand something you must master it. By constructing a finished product, the craftsman developed an intimate knowledge of it and a sense of control over it. Paracelsus felt that to understand ill-ness and well-being he could not be content, as the Aristotelian physicians in the university were, to analyze them according to first principles. He had to learn how to produce sickness and health so he could undertake proper treatment.

Paracelsus's desire to be in command of nature also drew him to the magic tradi-tion of his day. Natural magicians were indeed devoted to the control of nature. They undertook an aggressive program of experimentation, employing a host of means to gain power over the natural world. Theirs was, to be sure, a particular type of experi-mentation; they were more concerned with success than with why certain processes worked and others did not.

Paracelsus's willingness to bring the pragmatic orientations of both the craftsman and the magician to an academic subject—medicine—was an early manifestation of the desire to control nature in natural philosophy. It would grow more prominent in Britain in the early seventeenth century among those who, sympathetic to the practi-cal concerns of craftsmen and the heritage of natural magic, incorporated the goal of controlling nature into their natural philosophy.

Two New British Visions

William Gilbert.   One of the critics of established Aristotelian learning was a physician named William Gilbert (1544–1603). Like Paracelsus, Gilbert was open to the insights of the magic tradition of the Renaissance; indeed, Gilbert was aware of Paracelsus's alchemical writings. Standing squarely in the tradition of natural magic, Gilbert shared its deep commitment to experimentation as a means of inter-rogating nature. This was especially true, Gilbert said, when dealing with occult forces. Here the usual Scholastic dependence on reasoning from first principles was ineffective. There was a better way. When dealing with hidden or occult forces, "stronger reasons are obtained from sure experiments and demonstrated arguments than from probable conjectures and the opinions of philosophical speculators of the common sort."

The experimental approach he preferred was particularly evident in Gilbert's interest in and treatment of magnets. His work of 1600, *De magnete (On the Magnet),* contained a wealth of information about the magnet, which he obtained from the many experiments described in the book. At the same time Gilbert delighted to conduct experiments that dispelled legendary claims—for example, that garlic diminished the magnet's power.

But Gilbert had another motivation for writing his book on the magnet. He was a convinced Copernican and he wished to provide an explanation for why the Earth rotated on its axis. He had come to the conclusion that the Earth was a giant magnet. He also believed that magnets had the ability to turn spontaneously toward a pole. Because the Earth was a giant magnet, the Earth also had the power to turn spontaneously. But while the Earth's magnetism explained that the Earth could turn toward its pole, it did not explain why it kept turning, why it continually rotated on its axis.

Like Paracelsus and other natural magicians, Gilbert accepted that there were structural similarities that ran through the natural world at all levels. Through his work on magnets, Gilbert had become familiar with the work of the thirteenth-century physician Pierre de Maricourt, whose *Epistle on the Magnet* was first printed in 1558. In this work the author wrote that the magnet "bears in itself the similitude of the heavens." Gilbert used this similarity to draw a specific conclusion: just as the motions of the heavenly bodies showed them to possess an animate nature, so the magnetic motions of the Earth revealed it to be animate. The Earth, in other words, had a soul and that was what explained why it continued rotating every day. The point of the magnetic experiments involving the Earth was to show that the Earth had a soul. We see here in the work of William Gilbert how his commitment to experimentalism—a different approach from that of the Scholastics, who still dominated university life at the beginning of the seventeenth century—emerged from his participation in the tradition of natural magic.

The vision of Francis Bacon.   By far the most well-known natural philosopher of early-seventeenth-century England was Francis Bacon (1561–1626). Son of a highly placed courtier in service to Queen Elizabeth, Bacon was born to a life of wealth and privilege. Although he completed a degree in law, he practiced it very little. His real interest lay in a scheme he had first hit upon while still a student at Cambridge, the need for which had been reinforced during his travels in France.

His idea was to improve life in England by systematically revising the way the world was understood. As a student at the university, Bacon found himself thoroughly repelled by the Aristotelian philosophy he was taught. He failed to see how it related to the lives of English gentlemen. In France he had observed a level of refined culture that was not matched in England, but he had also observed corruption among the French. His new approach, he believed, would not only have practical relevance to English life, but it would also improve the moral well-being of the nation.

Bacon had begun to write up his grand idea soon after he was knighted. In 1605 an early version of the work appeared, called *Advancement of Learning*. The more complete account did not come out until 1620, appearing in Latin under the title *Novum organon (New Organon)*. Bacon chose that title as a way of announcing that he was replacing the *organon*, the term used at the time to refer to the corpus of Aristotle's logical writings. He had in mind nothing less than a thorough revision of Aristotelian natural philosophy; in fact, the *New Organon* was included as the second phase of a planned and uncompleted six-part work entitled *The Great Instauration*. The unusual term *instauration*, which he began using as early as 1603, but whose meaning he never specifically discussed, was intended to refer to the complete restoration or reconstruction of existing natural philosophy.

One of Bacon's criticisms of Aristotle's philosophy was that it contained no means of self-correction. He believed that the inevitable errors humankind made over time had disturbed what Bacon called the perfect "commerce between the mind of man and the nature of things" that humans once possessed. Bacon believed that Adam, the first man, had enjoyed this perfect relationship to nature, but that with the entrance of sin into the world human natural understanding had been corrupted. In the *New Organon* Bacon spelled out at length the kinds of errors—he called them idols—that had crept in.

The problem was that, left to the instruments of Aristotelian logic, these errors would go on forever. Bacon's conclusion was that there was but one course left: "To try the whole thing anew upon a better plan, and to commence a total reconstruction of sciences, arts, and all human knowledge, raised upon the proper foundations." As historian Steven Matthews explains, he saw this project in terms of Christian redemptive history. Christ's sacrifice not only provided a means of individual salvation; it also made it possible for humankind to restore a knowledge that led to God.

What were the proper foundations of knowledge? First of all, Bacon wanted to change the goal of natural philosophy from something contemplative to something active. With the entrance of sin into the world humankind had begun seeking knowledge for the wrong reasons. Natural philosophers had sometimes just wanted to satisfy their curiosity, at other times they wished merely to entertain themselves or even to build a reputation. But the real reason to seek an understanding of the world should be to benefit humankind by acquiring knowledge that was useful to everyone.

Another plank of the proper foundations of natural philosophy was the right method of procedure. Here Bacon took his cue from the inventions of the mechanical arts and crafts. History has shown us, he observed, that craftsmen learn from their experience; they improve the devices they create by building on past experience. As a result, knowledge of the mechanical arts was cumulative. It had not, as had the theoretical knowledge of nature, remained "almost in the same condition as it was, receiving no noticeable increase."

Bacon determined that an experimental approach, in which questions were put to nature that elicited a response, would permit the accumulation of practical wisdom. In this conviction he was greatly influenced by the traditions of alchemy and natural magic, with their commitment to experimentation. Like the alchemist who sought to uncover the natural sympathies and antipathies in matter, the magus strove to master nature's hidden powers and properties. In both cases the goal was practical knowledge that assisted the one who acquired it to exert control over nature.

Careful interaction with nature was the route to truth. Bacon claimed his method inverted the order of Aristotle's logic. The Scholastics jumped quickly from particulars to general premises. They then proceeded to the main concern of their analysis—teasing out specific implications of the general premises. But that would not necessarily lead back to nature. Bacon criticized those whose devotion to deductive reasoning made them adopt general premises too quickly. He claimed that his procedure emphasized the opposite direction—moving slowly and carefully from particulars to the universal (inductive reasoning). He employed induction to arrive at general conclusions about nature. By induction Bacon did not mean simply summarizing a series of individual cases—that was too simple and unreliable. He meant using individual cases to eliminate incorrect generalizations until there was but one generalization left. He proposed "to proceed regularly and gradually from one axiom to another, so that the most general are not reached until the last."

Bacon realized that an undertaking of the sort he envisioned must be done collectively. In 1626, the year Bacon died, he published a work of fiction called *The New Atlantis* that described an island where a project like the one he imagined was in place. It was a utopian portrait of what a rationally ordered society that was devoted to the acquisition of useful knowledge would be like. He described the various offices held within the society; for example, there were twelve "merchants of light" who traveled the world in disguise in search of books and experiments, three "miners" who selected and tried experiments, three "compilers" who summarized and classified the experiments, and three "interpreters of nature" who raised the discoveries from the experiments into more general axioms.

The Royal Society.   Bacon's ideas became important a generation after his death with the founding of the Royal Society. A small group of men, brought together in 1658 by a common interest in what was becoming known as "the experimental philosophy," met weekly either in the private home of one of their number or in a convenient tavern. Similar groups formed elsewhere, but the greatest activity was concentrated in the London meetings.

The group decided to include formal presentations and demonstration experiments in its meetings and to explore the founding of an organization "for the promoting of Physical-Mathematical experimental learning." By the summer of 1661 the members were trying to win the support of the crown. Charles II would not grant them funds, but on July 15, 1662, he did bestow a Charter of Incorporation on the new Royal Society of London. Reflecting its desire from the start to emphasize experimental philosophy, the charter provided for the appointment of two curators of experiments.

The members endorsed Bacon's views, especially his emphasis on the need to bring practical benefits to humankind through the study of nature. When in 1667 Thomas Sprat wrote *History of the Royal Society,* two figures flanked the bust of Charles II on the frontispiece of the book. One was the society's president, the other Francis Bacon, "Renewer of the Arts."

England was not the first country to establish a scientific society, but the Royal Society made its own distinctive mark on the early institutionalization of natural philosophy. The new curator of experiments, Robert Hooke, wrote in 1663 that the business of the society was to improve the knowledge of natural things and of useful arts and not to meddle in controversial matters like philosophy, politics, and religion. In spite of the declared intent not to discuss sensitive issues, however, the new society's loyalty to the religious and political establishment was beyond question. All was to be done "to advance the glory of God, the honor of the King, the Royal founder of the Society, the benefit of his Kingdom, and the general good of mankind."

◎ French Ideas on Matter and Motion ◎

Among Aristotelians, *organism* was the best metaphor by which to refer to nature. A major reason for this was that for Aristotle, natural processes—such as the actions of living things—occurred purposefully. In the natural world as Aristotle saw it even the four basic material substances endeavored to return to their own natural places when removed from them. They exhibited, in other words, behavior that reflected nature's larger purposeful structure.

The seventeenth century saw the appearance of a different view of matter and of the physical world. In this new conception the purposes evident in nature were not inherent in nature itself but were imposed from the outside. Here the metaphor for nature was *machine,* the workings of which accomplished the ends of its creator. To analyze nature involved acquiring knowledge of the motions of nature's moving parts. Matter in motion characterized natural process.

A New Conception of Matter

As we saw in Chapter 1, Aristotle explained the objects we apprehend through our senses as the union of form and matter. Attributes of form gave a material object its identity. The wood in a wood block supplied the matter, whereas the heaviness, square shape, hardness, and brown color were all attributes of form that gave the piece of wood its identity as a sensible object. Matter could not exist without form, nor could form exist without matter to serve as its subject.

As the seventeenth century got under way there were hints that another view of matter was emerging. In 1621 the French physician Sebastian Basso (dates unknown) published *Philosophia Naturalis (Natural Philosophy),* in which he abandoned the old Aristotelian conception of substance as the necessary union of form and matter. Basso regarded matter apart from form as something sufficient unto itself. It was permanent—fully capable on its own to establish the complete being of things that exist. Further, matter was incapable of undergoing change. It was dead, inert, and passive.

But if matter was unable to change by itself or be changed by anything else, how could there be any change at all in the cosmos? To answer this Basso and several others of his time appealed to change of place. Matter itself was wholly indifferent to change of place; that is, matter could change place without itself undergoing change. So if matter did change place, if it did move, it must be because motion was also a fundamental aspect of reality. While not required for matter to exist, motion nevertheless was always present.

A different conception of the eternality of motion from that of the ancients and even of Galileo was implied here and, as we will see, it soon emerged into natural philosophy. For the ancients and for Galileo, natural motion—that is, motion that would continue forever unless interrupted—was circular. The ancients believed that this kind of motion existed only in the heavens. Galileo suggested that it also existed below the orb of the moon. But the new conception of an ever-present motion resulted from matter's complete indifference to change of place. Matter did not care, so to speak, whether it was moving or at rest. If it was at rest, it would stay that way. If it was in motion, it would remain so. Nor did the motion in this new conception have to be circular. The next natural philosopher we consider, Descartes, argued that matter moving in a straight line would continue to do so until interrupted.

These two basic building blocks of reality—matter and motion—determined everything that existed and everything that happened in the physical world. To understand the world of nature, then, meant uncovering the particular kinds of matter and motion that characterized natural objects and their behavior. Clearly these ideas held profound implications for natural philosophy. In the first half of the seventeenth century they made their way into more than one interpretation.

Nature in Descartes's New Philosophy

The generation after Francis Bacon produced the most well-known transformation of natural philosophy based on the new ideas of matter and motion. The French natural philosopher René Descartes (1596–1650), son of a magistrate, rose from the modest beginnings of a military career to become one of the leading thinkers of Western history. As we will see, Descartes also opposed the Aristotelian Scholasticism of his day.

The influence of Isaac Beeckman. Descartes entered law school and completed a degree in canon and civil law in 1616. He decided to enlist in the army of the newly designated Prince of Orange in the Netherlands, Maurice of Nassau. Stationed in Breda, near the Belgian border, Descartes became acquainted with Isaac Beeckman (1588–1637), who had at that time just been examined successfully for a medical degree. Beeckman was interested in many things besides medicine; for example, he became familiar with the ideas of Sebastian Basso.

Among the ideas he communicated to the young Descartes was the notion that matter was made of tiny corpuscles whose shape and size, along with their motion, provided a means for explaining natural phenomena. Of course the ancient atomists had claimed something similar. What set Beeckman apart was his belief that the relationships involved here were capable of being described mathematically. In his view, mathematics was not restricted to the eternal motions of the heavens. It also could be used to capture the motions of the unseen corpuscles. According to Beeckman, Descartes became convinced of this "physico-mathematics" where natural philosophy was concerned.

The motive for a new method. Descartes soon left Breda, and his sojourn in the military, for several years of travel. Early on he had a dream that focused for him a task he then set for the future—to unify all knowledge into a system by creating a method.

He felt a need similar to the one that had motivated Francis Bacon—dissatisfaction with existing philosophy and the desire to create a new foundation on which to build a solid structure of knowledge. Both men saw in the Scholasticism of their day a system of knowledge that was out of touch with reality because it was erected on a foundation that lacked certainty. According to Bacon, Aristotelian thought could never produce certainty because it had no built-in capacity to correct error. Descartes came to a similar conclusion. There was much that could be doubted about Aristotle's worldview, but the Scholastics had no means of dealing with uncertainty in their understanding because the process of logical inference was regarded as above question.

Descartes received a sufficient amount of property and income from his parents to produce a satisfactory income for himself. As his thought matured he began to write down his ideas. But by the early 1630s, when he had sketched out much of his vision of natural philosophy, the condemnation of Galileo in Rome caused him concern as a fellow Roman Catholic. He withheld a work on cosmology, physics, and optics from publication, entitled *The World, a Treatise on Light,* because he suspected that its Copernican cosmology might offend. Later, especially in two works called *Discourse on Method* (1637) and *Meditations of First Philosophy* (1641) he spelled out his anti-Aristotelian ideas on the foundations of his new philosophy.

Descartes proposed to erect a firm foundation for knowledge by correcting the fatal flaw in Scholastic thought and in all other systems. He would reject everything that could be doubted, accepting for his foundation only that which could not be doubted. As optical illusions teach us, knowledge based on the senses clearly could be doubted, so his approach was clearly at odds with the inductive approach of Bacon.

René Descartes

Deductive systems were also fallible. We can deceive ourselves about the deductions of mathematics when we take as valid a set of inferences that later are shown to contain error. If both deductive and inductive reasoning could be distrusted, was there anything that was immune to doubt?

The process of doubting ended for Descartes when he tried to doubt his own existence. The very act of doubting was impossible unless he existed to do the doubting. Descartes's famous argument was summarized by the cogent conclusion, *cogito ergo sum* (I think therefore I am). Here was the basis on which to build.

Starting from this point, Descartes claimed to add only those propositions that shared the same degree of certainty. First, he knew he existed because he doubted it, so he must be imperfect as opposed to perfect (a perfect being would have no doubts). But the idea of perfection could not have been something he originated because it could not have come from something imperfect. So there must be a perfect being, God. Such a being would be less than perfect if he deceived us. When, therefore, we have ideas that we regard as clear and distinct, they must be true.

The nature of matter and spirit.   As noted earlier, one of the implications of Sebastian Basso's separation of matter and form was that matter by itself possessed the capability of establishing the complete being of things that exist. Another implication followed from that one—that mind or spirit, which in themselves had no material component, must be completely separate from matter. Basso and later Descartes accepted the fundamental reality of mind. It, too, was capable on its own of establishing being, but it was a different kind of being from material being.

Descartes further developed this idea, called metaphysical dualism, of two different basic kinds of reality. He argued that an essential difference between these two kinds of reality was that matter was "extended" being while mind was not. There were, he said, two kinds of things in the world—*res extensa* (extended things) and *res cogitans* (thinking things). Matter was the same thing as extension; it was, in other words, equivalent to space. Mind was not extended. Thoughts did not take up space. The world of material nature was constituted of extended things. God and angels were thinking things. Because humans possessed both a mind and a body, they participated in both worlds.

Natural things such as rocks, trees, and organisms (including other higher animals) were solely matter. Because there was no mind or spirit in nature, Descartes believed that he could not employ characteristics of mind when describing material nature. Nature was made of matter, which was dead, inert, and passive. That being so, nature itself and everything that happened in it must be the result of processes that were mindless. This is where the idea of a machine became helpful.

Machines, simply put, are matter in motion. The basic function of a machine is to transfer the motion of matter from one place to the next in order to accomplish the end of the machine's creator. One piece of moving matter impacts another, which presses on another, and so forth, until the last motion produced accomplishes the desired end. For example, a horse hitched to a millstone transfers the motion of its walking to the millstone, which then is used to grind wheat into flour. In such a circumstance force is communicated by contact from one place to another.

For Descartes, force is always transferred through the contact of one piece of matter with another and by no other means. By appealing to matter in motion, Descartes was able to satisfy the requirement that natural processes be mindless. A machine does not think or have purpose on its own, so the metaphor for nature in Descartes's natural philosophy became the machine. God had set the machinery of nature in motion at the Creation and it has been running ever since. All natural processes result solely from the transfer of that original motion from one place to another through material contact.

The mechanical cosmos.   Nature is full of instances where it is easy to observe a transfer of force by contact. Consider, for example, the erosive effect produced by moving water or the destruction of a house roof by a falling tree. But what about those instances where it does not appear that one piece of matter collides with or presses upon another and yet force is produced or matter is moved? What about the ball that falls after rolling off a tabletop, or the effect produced by a magnet on small pieces of iron? In these cases the cause of the motion has not been the impact of one piece of matter on another. How did Descartes explain them?

First and foremost, Descartes rejected the classical explanations of Aristotelian natural philosophy because they appealed to characteristics of mind. That is, Descartes eliminated the possibility of there being forces of attraction or repulsion, because attraction and repulsion were qualities of "thinking things." Any quality of mind or spirit just did not exist for him in nature itself. Aristotle's appeal to a rock "seeking" its natural place as an explanation of why a rock fell violated this strict separation of mind and matter, spirit and nature. The same rejection applied to the so-called occult or hidden forces such as the sympathies and antipathies referred to by alchemists and the psychical influences of astrologers. Categories like these applied to minds, but not to matter as Descartes understood it.

Recall that for Descartes, matter was simply spatial extension. We should not overlook two important implications of this understanding. First, matter (extension) did not possess sensory qualities such as color or taste. Color did not belong to matter. Color arose as an idea in our minds whenever certain motions of matter impacted our sensory apparatus. The idea of color was an arbitrary sign that nature used to help us make our way, just as nature had established laughter as a sign for joy. Nature could have used tears to communicate joy, but laughter was the sign instead. While color could theoretically have been associated with a range of matter's motions, nature used it as the sign for only certain motions. The second implication of matter's equivalence to extension was that matter's motion could be subjected to mathematical analysis. Descartes had first learned this from Beeckman, but here in his developed system it emerged quite naturally. Mathematics had long been used to measure and describe changes in spatial extension. Descartes's mechanical natural philosophy lent itself naturally to mathematical expression. His would be a mathematical mechanical philosophy.

Indeed, he came upon the idea of representing position in space by choosing a reference point and designating the position through three measurements from that point. From this grew the so-called Cartesian coordinate system, whose virtue was that geometric shapes could be represented by changing sets of numbers that defined a curve.

Descartes assumed that even where it did not appear that motion was the result of contact between two pieces of matter—as in the case of a falling object—it still had to be the case that such contact existed. There could not be such a thing as a vacuum because extended space (matter) existed everywhere. This was a major reason why Descartes opposed the ancient atomists, who had said that motion of atoms was only possible if there were a void, or empty space, through which the atoms could move. For Descartes, matter could only move as the result of contact with other matter, which filled the cosmos.

In the first part of *The World,* entitled "Treatise on Light," Descartes described three different elements of matter, distinguished by their size. Large masses such as the Earth and the planets constituted what he identified as the third element. The first element was the tiny fluid particles of no specific shape that emanated from the sun and exerted pressure that we perceive as light. The second element, formed as small globules, filled in around the large masses to fill the cosmos completely and transmit motion from one point to the next. Motion in any part of this arrangement could only occur if, as matter was moved, adjacent matter moved out of the way to accommodate it. That adjacent matter then must be accommodated by the matter it displaced, and so forth throughout the plenum of the cosmos. In order to prevent motion in one part of the system from producing motion everywhere throughout the whole, Descartes asserted that motion occurred in huge circular whirlpool vortices. Our solar system, for example, was but one such vortex.

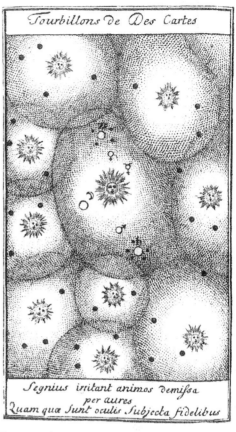

Descartes's Vortices

More than one physical arrangement might follow from the first principles he had established and Descartes was careful to acknowledge this. The matter that filled space could, through combinations of its elements, be arranged in a myriad of shapes and sizes. To see what particular physical circumstances exist in this world we have to observe the world. Then we figure out how they might be explained through the sizes, shapes, and motions of the parts. Descartes was not, as he is sometimes portrayed, a dogmatic rationalist who claimed to derive this particular view of the world out of pure ideas. More than one world could result from these motions.

Late in his life Descartes accepted an invitation from Queen Christina of Sweden, but within a few months he succumbed to an illness and died. His fame grew quickly, in part assisted by a certain notoriety he had acquired within the Catholic Church. His systematic doubt, his embrace of Copernicus, and his claim that deterministic mechanism ruled in nature all proved suspect to the authorities and his work was forbidden. But Descartes had erected a new system of natural philosophy that attracted many. His depiction of nature as a machine that could be described mathematically provided a practical means of explaining the natural world and it proved to be an extremely attractive alternative to the Scholastic Aristotelianism of the day.

The Revival of Atomism

Descartes appealed to the size, shape, and motions of the particles into which God had divided matter to explain such things as why oil was sticky, why salt tasted sharp, and why a piece of hot metal warmed the water into which it was put. In so doing he explicitly rejected the possibility of a void between particles of matter. A void could not exist, as matter was pure extension and there was no place in space that was not extended.

In rejecting the void, Descartes was voicing one of his objections to ancient atomism. As noted in Chapter 1, Epicurus, who lived from the late-fourth to the early-third century B.C., claimed that natural phenomena resulted from the collisions of pieces of matter moving through empty space. Certain aspects of Epicurean thought bore a similarity to that of Descartes. For example, both systems accepted the determinism of natural events as a result of the mechanical interaction of matter and for both systems such secondary qualities as taste and color existed only in the mind. But the absence of the void and a central role for God remained crucial, and fundamental differences existed between the two.

When, therefore, there appeared in France a revived version of Epicurean atomism by an almost exact contemporary of Descartes, it did so in opposition to Descartes's version of mechanical philosophy. Its author, the Frenchman Pierre Gassendi (1592–1655), had taken holy orders in the Roman Catholic Church and held positions as a philosopher and mathematician at various institutions of the church. What he tried to create was a Christianized interpretation of atomism.

Like Descartes, Gassendi developed an ardent opposition to Scholastic thought, especially Aristotelian natural philosophy. He had lectured on Aristotle's thought as a professor of philosophy, but grew increasingly dissatisfied with it and, in 1624, published *Paradoxical Exercises against Aristotle*. As someone gifted in mathematics and astronomy, he followed closely the results that were coming into public view during his early lifetime from Kepler and Galileo. He appreciated the critique of Aristotle he found in Galileo; in particular, he found himself drawn to the empirical and experimental approach he found there. His aversion to the dominant Aristotelianism of the universities was also bolstered by his readings of Christian humanists, including Erasmus (see Chapter 3), who appealed to ancient authorities in support of his critique of the dominant theology of his day.

Unlike Descartes, whose aversion to Scholasticism resulted in a questioning of all authority, Gassendi sought to replace Aristotle with another ancient authority. He found it in Epicurus, whose atomism he thought was well suited to the kind of empirical natural philosophy he endorsed. Emerging work with the new instrument of research, the microscope—which made its appearance around the turn of the seventeenth century—seemed to confirm the relevance of breaking objects into tiny parts for analysis. Gassendi believed, then, that natural phenomena resulted from the motions of atoms through the void of space. This was the foundation of natural phenomena—the particulars about their interaction had to be found by empirical research and experimentation.

In other words, Gassendi did not believe it was possible to learn which specific atomic motions produced given individual results in the natural world. His atomism, like his religion, was a position of faith. In fact, he distrusted the use of reason to deduce what was happening in nature. Reason could never penetrate to the knowledge of how the motion of atoms produced sensible results any more than it could uncover the truths of faith.

Not surprisingly, Gassendi believed that his vision of nature as a playground for the motions of atoms was not at all incompatible with his belief in God. Clearly, this was one aspect of the philosophy of Epicurus, for whom the gods were irrelevant, from which he dissented. Gassendi rejected rational proofs for God's existence, such as those that Descartes proposed. God was known through faith, not reason. Further, while Descartes's mechanical philosophy removed purpose from nature, Gassendi's did not. He saw final cause everywhere in nature and regarded it as evidence of God's intentions for and control of nature.

Gassendi and Descartes show us that among the new natural philosophies of the seventeenth century there could be substantial disagreement even between advocates of the same basic approach. We may regard the two as both mechanical philosophers, but they saw each other as adversaries. What was clear to all, however, was that the Aristotelianism found in the universities was encountering substantial opposition from many quarters.

◎ The Question of Materialism ◎

None of the natural philosophers of the seventeenth century that we have thus far examined questioned the existence of God. All accepted that God, who possessed the attributes assigned by traditional Christian theology, created and controlled the matter of the universe. God as spirit existed outside the physical cosmos, ruling as sovereign over it. The existence and nature of the cosmos depended directly on the will and power of the Deity.

The question arose among some, however, as to whether matter by itself could play the role that God held in traditional thought. If so, was there a function left for God as spirit to perform? As the century unfolded, those whose understanding resulted in sweeping declarations about matter's power elicited accusations of materialism—the belief that physical matter was the only reality and that everything, including thought, feeling, mind, and will, could be explained in terms of matter. This was the same charge that had been leveled against the ancient atomists. Materialism was seen as equivalent to atheism because, opponents declared, there was no role for God to play in the natural world.

Such accusations of godless materialism were not accepted by those natural philosophers who were exploring the extent of matter's scope. They felt, rather, that they were broadening the understanding of the divine. But, try as they might to explain themselves, they could not shake the charge of atheism. We will now examine the ideas of two such individuals from the seventeenth century.

The Monster from Malmsbury

When Thomas Hobbes (1588–1679) was still a boy, his father, who did not value education, left his son's schooling to his wealthy merchant brother, Thomas's uncle. From the age of eight he attended a school in Malmsbury, where his father had been a vicar. He then was sent to a private school, in which he mastered classical languages. Having demonstrated his keen intellectual abilities, he went off to Oxford at fourteen and completed his bachelor's degree in 1608.

Only in midlife did Hobbes become fascinated with mathematics and natural philosophy. Reputedly he came upon a copy of Euclid lying open in a library and on reading it became enthusiastic about the certainty present in mathematical calculation. He came to the conclusion that reason itself was nothing more than addition or subtraction or, as he put it later, "in what matter soever there is a place for addition and subtraction, there also is the place for reason; and where these have no place, there reason has nothing at all to do." While traveling to the Continent, Hobbes made the acquaintance of notable natural philosophers such as Galileo and Gassendi. He became enthusiastic about mechanical philosophy, though not about Descartes's dualistic approach that relegated it to the physical world only.

During his life Hobbes established close associations with the English aristocracy and the crown. As a result, he did not feel safe when civil war began to threaten in 1640. When revolution began in 1642, he was already in France. There he continued to associate with natural philosophers, even taking up the study of optics. He returned to England in 1651, the same year in which his most famous work, *Leviathan*, appeared. He took the title from the ancient monster of the Old Testament whose presence brought chaos in its wake.

In *Leviathan* he imagined how humankind would fare in a state of nature, without any form of government. In such a state chaos would rule; it would be a war of all against all. In order to avoid this, he explained, humans surrendered to a sovereign, who preserved order. Hobbes argued that in this arrangement the sovereign became a ruler whose authority was absolute.

As is evident from his blunt acknowledgment that humans submitted to authority in order to preserve themselves, Hobbes preferred to go straight to the bottom of things. The same was true for his understanding of nature. When he claimed that everything was either body or nothing, he was perceived to be denying that such things as spirit could exist at all. He was accused of materialism and atheism and regarded with suspicion everywhere in his homeland, despite the extensive discussion in his work of the word *spirit,* especially in connection to God. Whether he believed that God was material, or that God and spirit were simply not comprehensible, is still argued among philosophers.

In Chapter 9 of his book *On Body,* Hobbes confirmed his commitment to mechanical philosophy. He said that all change was "nothing other than the motion of the parts of the body undergoing change." Later he confirmed that the life of the mind was included among the changes brought about by matter in motion. Sensation was merely a motion of various internal parts existing inside the sentient being, while the exercise of the will was nothing more than the existence of appetite, a response to stimulus.

Like Descartes, Hobbes argued that there was no empty space in the universe. He specifically denied the possibility of a vacuum and became, as we will see, involved in a dispute with other natural philosophers as a result. Hobbes's natural philosophy, as that of the next figure we shall consider, showed that the study of nature and mathematics could produce results that confronted the very foundations of seventeenth-century Western society.

THE QUESTION OF MATERIALISM

Jewish Pantheism

Just as Hobbes's understanding of nature challenged the Christianity of his day, the natural philosophy of his much younger contemporary Baruch Spinoza (1632–1677) proved to be a thorn in the side of seventeenth-century European Judaism. Although he was born some forty-four years after Hobbes, Spinoza died two years before the author of *Leviathan*. But like Hobbes, Spinoza had to deal with the accusation of materialism and atheism from those in his religious community.

Spinoza was educated in the synagogue, where he studied the Talmud and classical Jewish authors. In order to learn Latin he worked with an independent tutor who was also a lover of the natural sciences. His tutor opened up a new world to Spinoza. He read Descartes and other non-Jewish writers and became enamored of their ideas. This contact with the ideas of European philosophers led Spinoza to question his religious heritage.

Like Descartes and Hobbes, Spinoza was impressed with the clarity and certainty of mathematical reasoning. He cast his major work, *Ethics,* according to the model of Euclid's *Elements.* That is, beginning with a set of definitions and self-evident axioms, he organized each of the parts of his book as a series of propositions that unfolded through deduction from one to the next.

Spinoza's philosophy centered on his definition of substance—a difficult but crucial concept in his thought. According to Definition 3, substance is "that which is in itself, and is conceived through itself." Spinoza elaborated by adding that substance was something whose conception was not dependent on any other thing but itself. God, for instance, he defined to be a substance with infinite attributes. It did not take the reader long to find out that, in fact, there was only one substance and it was God (Proposition 14).

Spinoza's viewpoint has been called pantheism—the position that identifies God with the universe. If everything that exists is God, then whatever is—minds, matter, motions—is an aspect of the one true being. Spinoza did not shy away from the implication that God did not act by free will. Why? Because the laws of nature, which operate with logical necessity as cause and effect, were merely aspects of a divine essence that was wholly self-contained. Hence, Proposition 32 stated that "Will cannot be called a free cause, but only a necessary cause." Everything that is and everything that happens must be as it is. God could not make decisions about what to do, but God (everything) could be considered free because he was not constrained by anything other than himself.

What Spinoza presented to the seventeenth century was a determinism that reigned not only in the physical world, as in Descartes's philosophy, but in the spiritual as well. He simply refused to break reality into matter and spirit—there was but one reality, substance. Of course it is easy to see why his ideas were so threatening to Judaism and Christianity alike. This was a radically different conception of God that did not agree at all with the orthodox theology of either religion. He was dismissed as another of the materialists and atheists that arose from an overvaluation of humankind's rational capacity to comprehend the natural world.

◎ The Mechanical Philosophy in Britain ◎

We have seen thus far a proliferation of new perceptions of the natural world in the seventeenth century. The appearance of Bacon's inductive empiricism, Descartes's rational mechanism, Gassendi's atomism, Hobbes's materialism, and Spinoza's pantheism were, to some extent at least, responses to new ideas that circulated widely in the first half of the century.

Facilitating the spread of information about new discoveries and new systems of thought were a few key individuals who positioned themselves at the center of correspondence networks of interested participants in the new science. Earlier in the century the French philosopher and theologian Marin Mersenne (1588–1648) filled this role, which was later undertaken by the astronomer Ismaël Boulliau (1605–1694) and the Englishman Henry Oldenburg (1618–1677). These individuals maintained contact with a host of scholars in Britain and throughout Europe, often bringing together disparate individuals who might otherwise have remained ignorant of their common interests.

As we have seen, the participants in the discussions came from every section of society. Some had to rise through the ranks to achieve positions of greater respect than that granted by birth, while others were well placed on the basis of their heritage. Among the latter was Robert Boyle (1627–1691), son of the Earl of Cork, one of the wealthiest men in Britain. Boyle would cite the work of Mersenne, Gassendi, Descartes, and others as key sources for his development of what he was the first to call "the mechanical philosophy." This British version, as we shall see, brought its own twist to the discussion.

Alchemy Continued

Scholars have recently argued that we cannot simply accept what Robert Boyle asserted regarding his intellectual inheritance. He downplayed, for example, the role of Daniel Sennert (1572–1637), a German Scholastic scholar who wrote about the merits of atomism even before Pierre Gassendi—its most well-known exponent in the seventeenth century—revived interest in Epicurus. Boyle's debt to Sennert, while undeniable, went unacknowledged because Boyle wished to associate only with those who were critiquing Aristotelian Scholasticism—not with one who identified with it.

But Boyle's choice of whom to acknowledge and whom to ignore involved an even more intriguing matter—the debt he owed to the continuing tradition of alchemy in the seventeenth century. Boyle wrote about the mechanical philosophy from his vantage point as someone immersed in the study of chemistry. In failing to acknowledge his debt to the alchemical tradition that first captured his attention, Boyle contributed to the notion that it was unimportant, not only to him, but to the development of ideas about how material substances combined.

Historians of science have become aware of the formative role played by a persistent and lively alchemical tradition in the seventeenth century. In Boyle's case, he learned a great deal from the American alchemist George Starkey (1628–1665), who wrote under the name Eirenaeus Philalethes and who was an enthusiast of the work of the Belgian mystical natural philosopher, Joan Baptista Van Helmont (1579–1644).

Boyle observed how Starkey valued number, weight, measurement, and experiment, which are usually regarded as the later hallmarks of chemistry. These emphases had been, however, central components of the alchemist's craft for a long time.

The Christian Chemist

As a young man of privilege, Boyle followed a stay at Eton College with extensive travels on the Continent. While in Europe he had a powerful religious conversion experience as the result of witnessing a severe thunderstorm. On his return to Britain in 1644 as a youth of seventeen, he settled into an estate left him by his father and decided to devote himself to the moral betterment of his fellow gentry. He pursued this goal through writing, including the composition of his "Occasional Reflections" on ordinary events and scriptural passages that were intended to inspire.

Among the subject matter Boyle encountered was that of the chemical laboratory, which piqued his interest so much that around 1649 his attention was drawn to it in a substantial way. He set up a laboratory in his house and began to construct experiments of his own. In 1651 he met Starkey, who by then was an experienced experimenter in residence in London, and the two became collaborators and correspondents. Starkey was in effect Boyle's tutor, instructing him on chemical and alchemical experimentation techniques during the formative period of Boyle's maturation as a natural philosopher.

Shortly thereafter, most likely between 1651 and 1653, Boyle composed a fragmentary piece he titled "Of the Atomicall Philosophy," a work heavily dependent on Sennert's thought. As mentioned above, Boyle did not credit Sennert, referring in the introduction to Descartes, Gassendi, and others who, he said, had "so luckily revived and so skillfully celebrated" atomism.

In 1655 Boyle moved to Oxford, where he became associated with a group of men who were enthusiasts for what was being called "the new philosophy" or "experimental philosophy." Around this time he seriously began reading, with the help of Robert Hooke, the writings of Descartes and Gassendi. His commitment to experiment, already reinforced by his association with Starkey, was further solidified through his acquaintance with Hooke and the others in the Oxford group.

Boyle embarked on an extensive program of experimentation and writing that set the tone for many discussions among British natural philosophers. In a work of 1660 he did experiments on the "spring of air," leading to the relationship between the pressure and volume that subsequently became known as Boyle's Law ($PV = k$). He called his particular version of mechanical philosophy corpuscularianism in order to avoid the irreligious connotations of materialism and atheism that atomism carried. Here again came the denunciation of the forms and qualities of Scholastic thought in favor of explanations based on the motion of matter, in his case the corpuscles that made up larger substances.

His embrace of mechanical philosophy was not so complete that its principles could explain all of reality. Corpuscles could be endowed, for example, with chemical principles that their motion did not explain. He entertained the possibility that there were what he identified as "cosmical qualities" that transcended purely mechanical laws. Nor did he abandon his interest in alchemy; indeed, his involvement with it reached a peak in the 1670s. Boyle had no difficulty acknowledging the reality of the nonmechanical results that alchemy could produce.

In addition to writing descriptions of his experiments, Boyle reflected on the larger meaning of his approach. Through such works as *Some Considerations Touching the Usefulness of Experimental Natural Philosophy* (1663), Boyle underscored the coming of a new age of science. Additionally, he used the results of the new experimental philosophy to support his strong commitment to religion. Beginning in the 1660s and lasting to the end of his life, Boyle wrote works that repeatedly explained how the study of nature led one inexorably to the idea of a Creator. As a distinguished member of the gentry, his position as a Christian gentleman lent credence to the claims he made as a natural philosopher. This was particularly helpful to him when he clashed with none other than Thomas Hobbes over the role of experiment.

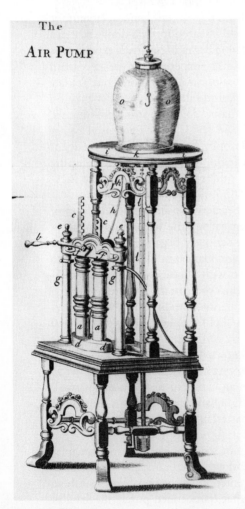

Boyle's Air Pump

Issues in Experimental Philosophy

One of the important factors that gave rise to the new experimental philosophy in the seventeenth century was the invention of new instruments for measuring and for magnifying. Instruments that magnified, like the telescope's improvement of our view of the surface of the moon, were not like new balances that merely improved an existing technique. Magnification intervened in nature—it gave the observer a view that was otherwise impossible to obtain through the normal use of the senses. It made possible what became known as "elaborate" experiments—testing (in a "laboratory") that carried the observer beyond the usual limits of our senses. This circumstance raised questions in the minds of some about the value of the information they provided.

In the late 1650s Boyle, with help from Robert Hooke, had an instrument built for him that produced a circumstance not normally met in nature. He called it a "pneumatical engine." It was an air pump, a device that would partially exhaust the air from a container. Using the air pump, Boyle conducted a series of experiments on the pressure of air and on the transmission of such things as light and magnetism through an allegedly evacuated space. Of course Boyle encountered all of the difficulties that accompany experimentation—the apparatus

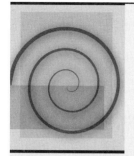

The Vulnerability of Experiment

The NATURE of SCIENCE

As the debate between Boyle and Hobbes reveals, experimentation in natural science cannot be the straightforward objective procedure it is sometimes portrayed to be. Differences in philosophical or religious outlook affect how an experimental procedure is understood. And many other factors condition the perceived meaning and significance of an experiment, not the least of which is the position in society of the persons evaluating it.

Robert Boyle was a Christian gentleman, a person of high social standing whose word and motives were not questioned by most people. However, due to the uncertainties involved in the air pump experiments, he felt it necessary to have credible witnesses confirm his results. This social conditioning enabled the witnesses to provide support for Boyle among the majority by introducing another level of credibility.

Thomas Hobbes, who was viewed as a materialist and an atheist, moved at the margins of accepted society and was excluded from the groups of natural philosophers who began gathering together around the middle of the seventeenth century. Hobbes in turn had no respect for immaterial factors such as the "spring of air" that Boyle relied on to explain his results. Such factors were dangerous deviations from the deductions about matter that geometrical reasoning inspired and that Hobbes believed formed the foundation of true philosophical knowledge. In opposing true knowledge, on which the absolute authority of the state rested, with allegedly neutral experiments, Boyle was putting the stability of society itself at risk.

The debate between Boyle and Hobbes illustrates well how philosophical, religious, social, and especially political factors can become involved in interpreting the results of experimentation. While the specific issues at stake will vary, such factors are rarely absent in any historical period, including our own.

was not perfect (in his case the pump leaked), results varied, and experimenters had to develop the ability to understand what they were seeing. There was plenty of room for objection among those who contested the process in the first place.

Thomas Hobbes was one of those who objected. When Boyle published an account of his experiments with the air pump in 1660, Hobbes responded the following year with his own treatise *against* experimentation. Hobbes opposed experimental philosophy because, he said, it was not really philosophy. Natural philosophers should explain by identifying the causes of things, not by experiment. Boyle would first have to explain what air was before he could gain any possible understanding of what was happening when he used his pneumatical engine. Because of the difficulties involved in doing the experiment, Hobbes was not at all persuaded that Boyle saw what he claimed to see. Boyle was deceiving himself. Hobbes offered alternative explanations of what was happening—the pump removed solid particles suspended in the air, but not the air itself because air, most likely, was infinitely divisible and would not be held captive by a piston.

The works of Boyle and Hobbes appeared just as Charles II was restored as English monarch after a period of revolution. Although Hobbes warned of renewed civil unrest if Boyle's new philosophy were allowed to take hold, Boyle represented the established way English society used to be and—now that the revolution was over—would be again. It was Boyle's view that prevailed.

◎ A Scientific Revolution? ◎

As historians of science just prior to the mid-twentieth century looked back to the period running roughly from 1450 to 1650, they were very impressed with the changes we have discussed in the last several chapters. What stood out clearly was the perception of various figures from those two centuries who saw themselves as engaged in something fundamentally new. Bacon, Galileo, and Kepler, for example, all used the word *new* in titles of their work. Because a number of the new attitudes would later become defining characteristics of natural science, the period was dubbed The Scientific Revolution.

More recently, however, some historians have suggested that it is misleading to refer to "The Scientific Revolution." First of all, there is no standard use of the phrase. Sometimes it is used to refer to events only in the seventeenth century while at others it means the period running all the way from Nicolas Copernicus to Isaac Newton, who will be the subject of our next chapter. But revolution usually is understood to occur in a relatively short time span, not 250 years. Further, the idea of revolution suggests that old ways were completely replaced by new ones. But, as we have seen, many of the innovations of the seventeenth century had roots running deep into the medieval period. Finally, what we mean today by the term *scientific* was a creation of the nineteenth century, so using *scientific* as if it applied to the early modern period misrepresents at least to some degree what was happening at that time.

Whether it is useful to designate a certain period of time from this era as The Scientific Revolution depends, then, on the purpose in view. There is no denying that, at least among natural philosophers, the conception of the natural world in 1650 was much different from what it had been in 1450. The change in conception can be regarded as revolutionary as long as we do not identify the period solely with the budding aspects of modern natural science that we recognize. As we have seen, there remained a rich variety of approaches to nature that did not generate a widespread consensus among the era's leading thinkers. That variety would continue for many years.

Suggestions for Reading

Peter Dear, *Revolutionizing the Sciences* (Princeton: Princeton University Press, 2001).
John Henry, *The Scientific Revolution and the Origins of Modern Science* (New York: Palgrave Macmillan, 2002).
William R. Newman and Lawrence M. Principe, *Alchemy Tried in the Fire* (Chicago: University of Chicago Press, 2002).
Steven Shapin and Simon Schaffer, *Leviathan and the Air Pump* (Princeton: Princeton University Press, 1989).

CHAPTER 8

───────────○───────────

Isaac Newton: A Highpoint of Scientific Change

Among the names of famous scientists that have remained widely recognized by the public, that of Isaac Newton ranks near the top. His exploits, especially the publication of his *Mathematical Principles of Natural Philosophy*, have inspired legends, many of which began in his own day. As he passed Newton on the streets of Cambridge, a student is reputed to have observed to his companion, "There goes a man that writt a book that neither he nor anybody else understands."

Almost everyone has heard about the famous apple, whose fall sparked an idea in Newton's mind that changed science forever. The idea was that all matter attracted all other matter and it made possible a mathematical description of the laws governing the motions of matter in the heavens and here on Earth. This achievement was an early step in the process of unification of nature's forces that has dominated physics ever since. Newton also made fundamental discoveries in mathematics and in the study of light and color.

Newton himself, in a moment of presumed modesty, claimed that if he had seen farther than others it was because he had "stood on the shoulders of giants." Indeed, Newton's accomplishments were only possible because of the work of the many natural philosophers who went before him. But there is also a sense in which Newton's brilliant integration of the disparate threads of prior individual achievements into a whole represented a plateau of scientific achievement that solidified the expression of a new worldview.

◎ The Background to Newton's Achievement ◎

Because Newton was not born into a life of privilege, the impression is sometimes given that he had to triumph over poverty in order to receive a university education and achieve what he did. Such was not the case.

Newton's Early Years

Isaac Newton was born early on Christmas Day in 1642, the first child from the marriage of Isaac Newton and Hannah Ayscough. At first glance, one would have little reason to predict that one of the world's greatest scientific minds would emerge from the match because no member of the Newton family before 1642 had had enough contact with formal education to sign his or her own name. That did not mean, however, that Newton's forebears were unsuccessful people. Isaac's grandfather was so prosperous a member of the yeomanry—the class of small farm owners below the gentry—that he was able to become lord of the manor of Woolsthorpe in 1623. This improved social rank also explains why his son, Isaac's father, married a woman from the gentry class, bringing additional wealth to the union. What the estate was worth became known only too soon, for Isaac's father died seven months after he married, leaving behind a pregnant wife.

Hannah Newton became the wife of the Reverend Barnabas Smith when Isaac was three, leaving him in the care of his maternal grandmother. It is clear from Newton's own confessions that the removal of his mother as his primary caretaker left a permanent mark on his development. His exceptional intellectual abilities made him all the more sensitive to the loss of her emotional support. Aware that he was unlike others because he had no father, Newton withdrew into himself. The guilt he felt for sins such as "punching my sister" was profound. He remained distant and mysterious to others and was known to virtually everyone as a difficult person.

As a small boy Isaac attended schools in nearby villages, but far more important than what he learned in school was the return of his mother to Woolsthorpe when he was nine. The Reverend Smith had died, leaving his mother with three new children, with whom Isaac now had to compete for her attention. Within two years Isaac was sent off to grammar school in Grantham, where he lodged with an apothecary who acquainted him with the fascinations of chemical composition, especially in medicines. Late in 1659 Newton's widowed mother called her 17-year-old son home from school to learn how to manage the estate, only to discover how unsuited he was for such a responsibility. On advice from her brother and the schoolmaster in Grantham, she consented to his return to school to prepare for the university.

Cambridge University

When Newton went off to the university he began an association with Cambridge that would last for the next thirty-five years. Naturally, what he accomplished during this time was in part due to the benefits that a university education and career offered him.

Trinity College.   In June of 1661, 18-year-old Isaac Newton entered Trinity College of Cambridge University. Obtaining a university education had changed in the preceding fifty years. Not that the curriculum had altered that much. The core was still Aristotle's thought, especially logic, and it, along with his physics and cosmology, were among the formal subjects Newton encountered early. But studying Aristotle hardly represented the cutting edge of European thought; in fact, students at Cambridge learned mainly through the rote mastering of texts that generated little if any intellectual passion. Apparently, the major justification for retaining the curriculum was simply that it had always been that way.

What had changed at Cambridge was the manner in which the college system operated. Whereas earlier tutors had taken their responsibilities to younger pupils seriously, by the second half of the century many fellows simply took their stipends without concerning themselves with their younger charges, who were expected to fend for themselves. Newton began to go his own way quickly, even though his tutor suggested a standard course of reading assignments. It must have become obvious to Newton that he possessed abilities in excess of those around him, including some of his professors.

The impact of self-instruction.   The new direction Newton began to follow early in 1664 resulted from his reading of works by Descartes, Gassendi, Galileo, Hobbes, and others. His procedure was to record in a special notebook the questions that these authors provoked as he read their explanations of natural phenomena, even suggesting possible tests by which his questions might be answered. Newton was drawn more to Descartes's mechanistic explanations—according to which natural phenomena were explained as the result of impacts between kinds of matter—than to the so-called qualities of matter Aristotle had designed precisely to fit the explanation sought. But Descartes's system also worried him. At first, Descartes's philosophy appeared to be friendly to religion, but Newton began to ask whether Descartes's confinement of spirit to the realm of "thinking things" meant that God had been excluded from nature.

Newton bought Descartes's book on geometry just before Christmas of 1664, although he had already read it six months earlier. Mathematics captured his attention, even though he had no real background in it and had to teach himself for the most part. Mathematics was regarded as an esoteric subject in the university. If he wished to win election as an undergraduate to one of the sixty-two scholarships the college controlled, studying mathematics was not a good way to distinguish himself. But Newton was in fact elected, some assume because of the support of someone who was well placed in the structure of Trinity College who intervened on his behalf.

◎ Newton's Central Interests ◎

Newton spent the next four years preparing for his master's degree, which, when he obtained it, made him a permanent resident of the university as a fellow of Trinity College. However much Newton depended on those who had preceded him, clearly he was one of a kind at Cambridge. He stayed mainly to himself, making few friends. One exception to this pattern of isolation was his attendance at the lectures of Isaac Barrow (1630–1677), the Lucasian professor of mathematics. Barrow had learned enough about Newton's mathematical abilities during the years when Newton was preparing for his degree to be highly impressed. When Barrow resigned his position in 1669, he recommended that the young master of arts succeed him in the Lucasian chair. In the fall of 1669 twenty-six-year-old Isaac Newton landed a professorship, his future all but assured.

Mathematics and the Theory of Color

Newton brought innovations to two subjects that early captured his attention. Even before he assumed the Lucasian chair of mathematics, he embarked on the creation of a new approach to the study of things that change continuously with respect to each other, now known as the calculus. Inventing a means of representing such change, whether the rate of change was uniform or not, opened up new possibilities for the mathematical depiction of natural processes. In the study of light, Newton combined his talent for experimentation with his acute power of reasoning to produce a new understanding of the nature of color.

Newton and the calculus. Newton came across a work by John Wallis (1616–1703), in which the Oxford geometer had used so-called infinitesimals in calculating the area under certain curves. The idea of infinitesimally small quantities, in particular an infinitesimally small piece of a line, dated back at least to the Middle Ages. Wallis used this idea to compare the area of certain curved geometrical figures to that of a square.

Newton took Wallis's work one step further. Rather than conceiving of a curve or line as being composed of infinitesimally small pieces statically joined together, Newton began to think of curves as being generated by motion. The idea of motion entails, of course, change over time, so it was natural for Newton to wonder about what he called the "moments" of change. Curves and lines express how one varying quantity (the dependent variable) changes in response to a change in the other (the independent variable). In straight lines the relationship of change is uncomplicated. Newton now employed the infinitesimal to express the moment of change in a *curved* line at a given instant of time. He did this by considering the ratio of the infinitesimally small change of the dependent variable to that of the independent variable, a ratio of what he called *fluxions*.

The "infinitely little lines" of each variable, as he called them, were both real and not real because an instant of time represents an interlude of no duration. Somewhat later Newton's fellow countryman George Berkeley criticized him by observing that his "evanescent increments" appeared both to exist and not exist, adding, "May we not call them the ghosts of departed quantities?" But Newton treated them as real, manipulating them as he did other algebraic entities. In so doing, he had invented a new and powerful means of analyzing and solving highly complex mathematical problems. Time would show that Newton's invention also would have a huge impact on the solutions to the problems of everyday life.

In October of 1666, he wrote a tract on resolving problems by motion. There is no indication that he showed it to anyone; nor did another general work on the analysis of equations from 1669 make its way into public view. A few individuals saw the second tract, but perhaps because it did not make any explicit or pointed use of the fluxional notation, no one deciphered from it the scope of Newton's abilities.

Newton did later publish his method of fluxions as an appendix to his famous book, *Mathematical Principles of Natural Philosophy.* Eventually there erupted a protracted disagreement about who invented the calculus, with bitter charges flying back and forth across the English Channel. Gottfried Wilhelm Leibniz (1646–1716), a German philosopher, mathematician, and diplomat who had seen a private copy of Newton's 1669 work on equations in 1676, assumed at the time that Newton's mention of infinitesimals contained nothing different from what others, including Wallis, had done. The year before, Leibniz had invented what amounted to an alternative form of Newton's method of fluxions. In the question of who invented the calculus, then, there were clearly two winners and no losers, since scholars agree that both men created the same mathematical tool independently.

The theory of color. For his first course of lectures as a new professor at the beginning of 1670, Newton chose a subject he had been investigating before his appointment—the nature of colored light. Newton described experiments he had performed with prisms that suggested a new understanding of the prism's effect on a beam of light. He rejected the common explanation that the prism separated white light into colored light by weakening it somehow. In that explanation the red light (and all other colored light) that emerged from the prism was really just the white light that had been forced by the prism to appear red.

Newton concluded instead that colored light constituted the white light's component parts. Colored light was more basic than white light because the latter resulted from combining all the various forms of the former. Newton came to this conclusion using what he called a crucial experiment, by which he meant an experiment that was able to settle the matter once and for all.

When he let a beam of white light enter a prism, it was broken into the expected colored beams. By rotating the prism he was able to select one of the colored beams, for example red, so that it passed through slits and entered a second prism. Here he noticed that the beam of red light was not changed by the second prism as it passed through it. Newton noted that the second prism had not weakened the red beam, as it should have under the usual explanation. What *was* the same, he observed, was the angle through which the light had been bent. The angle formed by the incident beam of white light and the emerging red light in the first prism (angle A in the figure) was the same as that between the entering red light and the existing red light of the second prism (angle B).

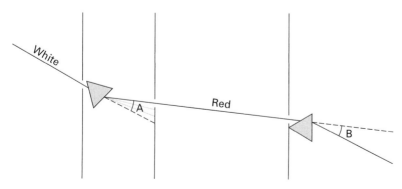

Newton's "Crucial Experiment"

Furthermore, he saw that when he selected a different beam of colored light to pass through the second prism, the size of the angle through which the light was bent was different from that of red, but the angles in the two prisms were equal— just as they had been in the case of the red light. He concluded that he had established that the beams of colored light, far from being a modification of white light, actually were the component parts of white light, and that the angle through which a particular colored light was bent was its defining feature.

As part of his studies, Newton built a reflecting telescope of which he was quite proud. When in 1671 members of the Royal Society in London asked to see it, Newton sent it to them and was promptly elected to full membership in the society. Early in 1672 he sent them a paper summarizing the several years of work he had done on color. As a result of the telescope and the impressive paper on color, which was immediately published by the Royal Society, Newton's name began to be mentioned by those living outside Cambridge.

Newton's work on color aroused opposition from Robert Hooke, a defender of the modification theory of color, and Christian Huygens, who objected that, by only appealing to the merely accidental characteristic of colored light's refrangibility, Newton had never really explained what color was. Newton's responses made clear that he was a testy, if gifted, thinker who would tolerate little criticism. His overreaction was childish, indicating, at the very least, an extremely eccentric personality. Newton simply withdrew, intending, as he put it, "to be no further solicitous about matters of philosophy." But his understanding that colored light formed the basic components out of which white light was composed became the dominant view.

Alchemy and Theology

After this brief contact with the scientific world beyond Cambridge little was heard from Newton for over ten years. He turned to work he had already begun on the motions of heavenly objects and to his serious interest in both theology and alchemy. His treatment of these latter two subjects, both of which concern themselves with the relationship between matter and spirit, indicates Newton's growing resistance to Descartes's mechanical philosophy. In Descartes's approach spirit is by definition excluded from nature (see Chapter 7). Newton was, in fact, close to rejecting the Cartesian assumption that the only way one body can exert force on another was by means of an impact between the two, as occurs in the working of a machine. If spirit were somehow present in the physical world, might not it mean that this presence would be reflected in the interactions of matter?

The centrality of alchemy in Newton's thought.   Because of his work on colored light Newton began to receive letters from people wanting to know his thoughts on a variety of subjects. Repeatedly, Newton would excuse himself from giving a substantial reply with the explanation that he was heavily engaged in some private business that was taking up his time and demanded his complete attention. It was not mathematics and it was not light and color. Many agree that the it most likely was Newton's study of alchemy.

From at least his grammar school days Newton had harbored a curiosity about chemistry. Sometime around 1666 he began to organize in his mind how one might go about studying the way in which material substances interact. One of the first authorities he consulted was Robert Boyle (see Chapter 7), in particular Boyle's book on forms and qualities. Newton learned about furnaces and their operation, an unmistakable clue to his intention of doing chemical experiments for himself. Before long it became clear that his real interest in the subject, if it had not been from the start, was alchemy.

Newton undertook the study of alchemy in the same systematic way he had investigated the theory of color. His procedure was to set down axioms to govern his investigations and to compile notes for and from experiments. He assumed that whatever changes he observed involved a chemical process rather than some kind of mystical transformation; hence he determined to take exact physical measurements, as he had in his optical experimentation. Newton disdained the appeal that alchemy held for those he saw as "ignorant vulgars," who sought how to become rich by learning how to turn lead into gold.

Newton always remained convinced that particles of matter in motion constituted reality. But never did he believe that a description of matter in motion supplied a complete description of reality. That view was too narrow and restricted, and it did not presume to explain the nobler motions of matter we humans encounter. To understand a material entity, Newton was not satisfied with hypothetical corpuscles whose only characteristic was extension. He took his cue from the human body because it also possessed a mind. Unlike Descartes, who separated mind and body, Newton saw them as united. The manner in which the human mind related to the body—a mental decision to move one's arm could result in a physical movement—was a more complete model for understanding why matter moved than the mere mechanical analysis of impact.

Newton suggested that the motions of matter controlled by the human mind provided a good means of imagining how the motions of the heavens were subject to God's control. In both cases passive matter was animated by an active agency. Newton was thus able to claim that an action did not have to rely on an intervening medium to be transmitted and at the same time avoid attributing the action to powers belonging to matter itself. This way of understanding matter and its forces was quite compatible with the view of matter in alchemy. A fundamental alchemical conviction, for example, was that material nature was infused with the active principles of the feminine and the masculine in whose union generation of some kind proceeded. Newton agreed. He felt that if we could understand which substances possessed which principles it might be possible to find out what combinations generate new substances. Newton also spoke of what he called "the principles of vegetable actions," which he contrasted with mechanical actions. These were nature's agents, "her fire, her soul, her life," and they involved the presence of a subtle kind of matter that he thought of as diffused throughout the "grosser matter." This new kind of matter, an ethereal substance, was the bearer of the active agency, for "if it were separated there would remain but a dead and inactive earth."

Animating force lay at the heart of Newton's thought. He speculated about other kinds of ethereal substance that produced activity in nature. One he suggested might be involved in the propagation of light. Other "ethereal spirits," he said, might be involved in the production of electrical and magnetic phenomena and perhaps in the gravitating principle. The idea of an animating force, then, links his interests in alchemy to his conception of nature's forces in general, including his view of gravity. He let his mind freely wonder if maybe all of nature might be nothing more than the precipitates of "certain ethereal spirits" that God initially shaped into various forms and that had been molded "ever since by the power of nature."

A theological crisis. Among the interests seething in Newton's mind in the 1670s was theology; in fact, theology forced itself on him with an urgency that pushed alchemy into the background for a brief time. Newton had always approached theology with the same seriousness he devoted to all his studies. His careful examination of the Bible convinced him that the Scriptures had been corrupted. He believed that writers in the fourth century had in fact altered original texts in order to promote the divinity of Jesus. Newton's theology agreed with that of Arius, the fourth-century Alexandrian priest who taught that Jesus was not coequal with God the Father but was inferior to him, having been created and then elevated to his right hand. This unitarian view threatened his future at Cambridge in a very pressing manner.

As a master of arts and a fellow of Trinity College, Newton was required to be ordained to the Anglican clergy or be expelled from the college. With the deadline for his ordination approaching in 1675, Newton knew his private nontrinitarian views were wholly incompatible with ordination. He asked to be excused from the requirement, but since he did not think his request would be granted, he prepared to leave Cambridge. Then a special dispensation from the crown exempted the holder of the Lucasian chair from the requirement of ordination. No one knows for sure why the dispensation was granted, although it is unlikely that it was intended to permit Newton to retain his secret unitarian beliefs. It is more probable that Newton and all future Lucasian professors, were, as mathematicians, deemed unsuitable candidates for the Anglican ministry.

The Motions of the Heavens

In 1687 one of the most famous books in the history of science appeared in England. It was a sensation that made Newton an undisputed authority in natural philosophy. *Mathematical Principles of Natural Philosophy* (known widely since then as the *Principia*), revealed its author, still largely unknown to many of his day, to be among the world's most gifted thinkers. Readers quickly discovered that Newton was aware of all the important recent achievements of natural philosophers in Britain and on the Continent. Further, Newton had clearly thought carefully and in great detail about the nature of matter and motion and he had supplied impressive new mathematical treatments of the complex motions of the heavens that men such as Kepler and others had struggled with before him. The book may have seemed to many to burst forth out of nowhere, but Newton had in fact been studying natural philosophy and the motions of the heavens for some time.

The problem of the moon's motion.   Newton agreed with Descartes that natural motion, or motion that did not require a mover, occurred in a straight line and not in a circle as Galileo believed. But if that were so, then something must be preventing the moon from moving in a straight line by bending its motion into an orbit around the Earth. It was as if a cosmic string of some sort attached the moon to the Earth, pulling it into a circular path around the Earth much as a rock could be twirled on a string. But what was this cosmic string?

The well-known story about the fall of an apple did not come directly from Newton. An acquaintance of Newton named William Stuckley reported that on April 15, 1726, the year before Newton died, the two of them dined together and then went into the garden to drink tea under the shade of some apple trees. Newton told Stuckley that he had been in the same situation as a young man when the fall of an apple triggered an idea in his mind about how to solve a problem that had been puzzling him. The problem had to do with the moon's motion. But Newton's work with the problem of the moon extended over a much longer period than many have assumed. Only gradually did he become clear about what the complicated motion of the moon entailed, some aspects of which required almost two decades to resolve.

Falling apples and a falling moon.   When the young Isaac Newton casually saw an apple fall from a tree during his stay at home in Woolsthorpe, it apparently occurred to him to consider more carefully the reason *why* it fell. Aristotle's idea of natural place, according to which the heavy elements earth and water move toward the center of the cosmos, no longer satisfied him. He had come to regard that explanation as begging the question. He preferred to think of the apple as subject to a force that caused it to fall, a force that was somehow associated with an object's heaviness, or what was known as *gravitas*.

Part of Newton's insight was to recognize that he did not need to solve the problem of how the force was applied in order to understand what made the moon bend into an orbit around the Earth. What if the force, regardless of how it worked, affected the moon in the same way it affected apples? The biggest drawback to this line of thinking was that the fall of apples occurs at the surface of the Earth, while the moon is a very long distance away. Newton reasoned that because things like apples continue to fall even at the tops of high mountains, perhaps the force acting on them extends much farther from the center of the Earth than people normally had assumed. Maybe it even extended all the way out to the moon. If that were true, then could the moon, like apples, be considered a falling body?

Much later, after he had figured out his solution to the problem of the moon's motion, Newton published a diagram (see artist's rendition on next page) to explain why one might, in fact, regard the moon as a falling body. He imagined throwing an object like an apple from a mountaintop located at the top of the Earth. (See the figure.) If one threw the object along the tangent line VP in the diagram, then when the object was released it would undergo natural straight-line motion unless acted on by

The Moon as a Falling Body

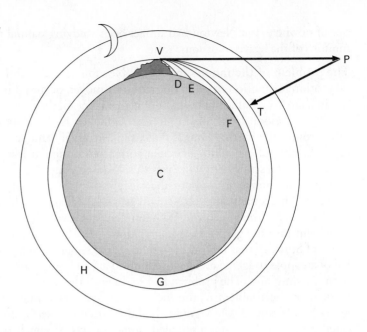

some force making it act otherwise. But *gravitas* affects objects on mountaintops, so the object would fall toward the Earth's center at the same time it experienced natural straight-line motion along the tangent line. The combination of the two motions would result in a curved path leading from the mountaintop to the Earth's surface. Newton depicted several different paths of fall to the Earth's surface (VD, VE, VF, VG), each depending on how hard the object had been thrown along the tangent line.

Newton's diagram also depicted how the object could be thrown so that it orbited the Earth and yet could be still considered on one of the paths of "fall" to the Earth (VH). In the case of a satellite, however, the object falls without ever getting closer to the Earth's surface. One can understand this phenomenon by separating the curved motion of the path the object traverses into the two simultaneous motions that make it up. If the object is thrown just hard enough along the tangent line from the mountaintop, it would travel away from the Earth far enough in any unit of time (from V to P in the figure) so that the distance it falls back toward the Earth *in that same time* (from P to T) would place it at the same distance above the Earth as when the object was thrown. The moon does in fact orbit the Earth, so could one not conclude that it "falls" toward the Earth like any other falling object? Newton had in fact grasped a way in which the moon might be like the apple he had seen fall in the garden. If he was right, then the moon, too, was subject to *gravitas*.

Learning a lesson from Hooke. As he thought more about the motion of objects, one particular challenge emerged. As noted, Newton agreed with Descartes, who had died in 1650, that the motion objects would undergo if left to themselves was uniform motion in a straight line. But there was also a difference between the views of Descartes and Newton. Descartes held that an object undergoing such natural motion was not being acted on by a force. Newton, on the other hand, thought that natural motion resulted from the action of an active force that matter possessed "by which any being endeavors to continue in its state and opposes resistance."

Newton was not content, as was Descartes, to confine the action of force to the impact between two masses. He acknowledged that force was applied when one piece or one kind of matter struck another, but he could not agree that this was the only way in which force was exerted. Newton's understanding of matter as a passive mass that was animated by an active principle was as far from mechanical philosophy as it was close to alchemy. He was certain that God was the ultimate source of this active principle. "God who gave animals self motion beyond our understanding," he wrote in 1675, "is without doubt able to implant other principles of motion in bodies which we may understand as little."

Newton apparently did not realize that his conception of active force had produced an inconsistency in his thinking. Like Descartes he had referred to "the moon's endeavor to recede from the earth," but unlike Descartes he assumed that this tendency was due to an active force present in the moon's matter that caused uniform straight-line motion. Circular motion he saw as the balance between two equal but opposite active forces, one toward the center of motion (causing, in the case of the moon, its "fall"), and the other away from the center (due to the moon's "endeavor to recede"). In 1679 he received a letter that woke him up to what was wrong with this assumption. And worst of all, the letter was from an old critic of his theory of color, Robert Hooke.

Hooke, who was secretary of the Royal Society, tried to draw Newton into a discussion of ideas on the system of the world. Five years earlier in 1675, Hooke had claimed that the motions of the planets around the sun could be explained as a combination of their natural tangential motion with "an attractive motion toward the central body." This was clearly too close to work Newton had been considering for a long time for Newton to ignore. He had been thinking about such matters ever since he had seen the apple fall as a young man in his early twenties. Hooke's renewed contact had the effect of drawing him out again into an exchange of views. It would also force Newton to face squarely the implications of his ideas of force.

In the ensuing correspondence it became clear to Newton that Hooke had understood something Newton had not. He may have hated to admit it, but Hooke's explanation of planetary motion avoided the mistake he had made in thinking of circular motion as involving both inward and outward forces. Hooke explained the motion as the result of the planet's tangential motion, combined with an attractive motion toward a central body, *making no reference at all to a force that made the moon tend to recede from the Earth*. Newton now realized that a consistent account of the cosmic string holding the moon in its orbit needed only natural straight-line motion and the presence of a force that pulled the moon *out* of that motion into an orbit around the Earth.

Edmond Halley's famous visit.   One day in August of 1684, some five years after his exchange with Hooke, Newton received a visit from the astronomer Edmond Halley. Halley knew of Newton's mathematical gifts, and, finding himself in Cambridge, decided to consult with him about a problem on the minds of several people in London. Halley knew of Hooke's work on the system of the world and he was aware that it included a specific conjecture about the attractive force acting toward the center. Hooke had even suggested a mathematical formula for calculating how strong the attraction was at various distances from the center. Halley himself had confirmed that if one started with a circular orbit, one could show that it was governed by Hooke's formula.

The London group suspected that if the formula was correct, then it should be the foundation for the laws governing celestial motion. In fact, theoretically it should be possible to show the converse of Halley's demonstration; that is, start with Hooke's formula and derive the orbits of the planets from the force law that determined them. Going in this direction, however, was much more difficult; further, everyone knew that the orbits were not circles but ellipses.

Halley put the question straight to Newton: if the planets were drawn toward the sun according to the formula, what path would such a formula entail? In one account Newton is said to have answered immediately that the formula entailed an ellipse for the planets. Halley was astonished that he answered so quickly and confidently. "How do you know that?" he asked. "Because I have calculated it," Newton reputedly replied. When Halley requested to see the calculation, Newton claimed to have misplaced it, but he promised to send it to him.

Few Newton scholars take him at his word about the lost calculation. Some suggest he simply did not want to disclose it hastily for fear of what Hooke and others might do with it. Others wonder if Newton had in fact completed the calculation by that point. In any event, by November of 1684, he sent Halley not only the solution to the problem, but much more as well— a tract of nine pages on the motion of bodies in orbit that contained several general conclusions with broad implications. For example, Newton showed that Hooke's formula entailed an orbit that is a conic section, which is elliptical when the velocities are slow enough, as they are for planets.

Newton's proof that gravity affects the moon.   Halley's question to Newton in 1684 touched on a subject Newton had in fact thought about over many years, ever since he had first considered the problem of the moon's motion. Back in the mid-1660s, when he was at home because plague had forced the university to close, he had devised a strategy for testing whether gravity affected the moon. It involved using Galileo's figure for the diameter of the Earth, which was in fact not very accurate. The results he had gotten in 1666 were good enough to convince him that he was on the right track. But they were not close enough to what he had predicted they should be—based on his own beliefs and on his knowledge of Descartes, Kepler, and Galileo—to satisfy him that he had completely solved the problem. The proof Newton developed in the wake of Halley's visit used the same strategy, this time with more accurate data, and it became part of his famous book on natural philosophy.

The strategy for testing whether gravity affected the moon came out of a reply Newton gave to a criticism of Copernican astronomy. Like Galileo and Kepler before him, Newton was a Copernican. Opponents of Copernicus had argued that an Earth rotating on its axis would fling objects on the surface off into space. Newton showed that the heaviness of objects supplied a much greater force than that outward tendency generated by a spinning Earth. Shifting his focus from a rotating Earth to the moon rotating around the Earth, he compared the tendency of the moon to be flung out of its orbit by the force of its revolution around the Earth to the gravitas that made it a falling object.

In Book 3 of the *Principia* of 1687, Newton explained how one might prove that the force that made apples fall also acted on the moon. He established his conjecture that the apple force supplied the cosmic string he sought by using a clever line of reasoning. First he *assumed* he was right about the apple force also affecting the moon. By combining that assumption with other knowledge he had about how objects move he realized he had enough information to make a prediction about how far toward the Earth the moon would "fall" in one minute. He then could check to see if his prediction could be confirmed. If it could, then his assumption must have been correct. First, we must understand the foundation for his prediction, then the prediction itself, and finally the confirmation.

1. *The foundation for the prediction.* Before Newton could make a prediction that he could test, he had to find a formula that expressed how the apple force varied over distance. He suspected that the farther away apples were, the weaker the gravitational force he believed was pulling them toward the Earth. But he had to find out precisely how the attractive force varied. The relationship he discovered has become known as Newton's inverse square law. It is the same formula that Hooke suggested in the system of the world he proposed nearly a decade after Newton first came to it in 1666.

Without recreating the exact route to the inverse square law, suffice it to say that Newton was able to deduce it by building on earlier conclusions. He began with his claim that a body would naturally continue in uniform straight-line motion forever unless interrupted. An obvious implication of this conviction was that any motion that was not natural, such as curved motion or motion that was otherwise accelerated, was being caused by a force. Newton, in other words, had learned to associate force and acceleration. For Newton, then, $f \propto a$ and $a \propto f$.

From here Newton drew on other things, such as Kepler's third law, to conclude that the apple force (and therefore also the acceleration of a falling apple) would diminish according to the formulas $f \propto 1/r^2$ and $a \propto 1/r^2$, where r is the distance the apple is from the Earth. This is Newton's famous inverse square law and its discovery served as the foundation for a prediction he could verify.

2. *The prediction.* Newton's prediction was that if the moon was affected by the same force that makes apples fall, then the distance an apple would fall at the Earth's surface in one minute was 3,600 times greater than the distance the moon fell in the same amount of time. He was able to calculate this prediction so precisely because the moon was known to be 60 Earth radii away, or 60 times farther away from the Earth's center than falling apples were. So, if whatever made apples fall also made the moon "fall," then the rate of acceleration of the moon's fall would, by his inverse square law, be weaker by a factor of $1/60^2$. The assumption that gravity affected both apples and the moon also meant that the distance the moon "fell" in a given amount of time would be $1/60^2$ times less than the distance apples would fall in the same amount of time.

Using an argument from geometry and his knowledge about the moon's distance from the Earth, Newton calculated that the moon actually "fell" 15.083 Paris feet (approximately 16 English feet) toward the Earth in one minute. If his assumption

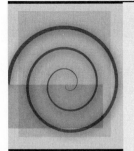

The NATURE of SCIENCE

The Status of Newton's Proof

Newton's inference that gravity holds the moon in its orbit is an example of scientific proof. But exactly what kind of reasoning is involved? Consider the statements p and q below:

p: the same force that makes apples fall also makes the moon "fall"

q: apples on Earth fall 3,600 times farther than the moon in the same amount of time

Newton's argument is:

If p is true then q is true.

q is true.

_____

Therefore p is true.

As impressive as Newton's demonstration is, its logic does not always hold true. There are many examples of this form of logical argument that are untrue. For example, let p be the statement that I won the lotto and q be the statement that my taxes increase. The previous argument then runs:

If I win the lotto then my taxes will increase.

My taxes have increased.

_____

Therefore I won the lotto.

Most scientific proof uses an argument of this structure. We conjecture that something causes a certain result and then seek evidence that the prediction is true. Finding the evidence, we conclude that the conjecture has been verified. Of course the persuasiveness of the proof depends on various factors, such as how exclusive the proposed cause appears to be or how improbable or precise the prediction seems. Because other factors may be involved in the production of the predicted result, we can never be certain beyond all doubt that the identified cause is exclusively responsible for the result. In fact, even the conceptual and linguistic categories we use when forming a hypothesis can carry with them hidden assumptions that affect the meaning of the result and that only become clear to later generations or people from different cultures. So the permanent acceptance of the cause is never guaranteed.

that the acceleration of the moon's fall was 60^2 times weaker than at the Earth's surface were correct, that would mean that objects near the Earth should fall $60^2 \times$ 15.083 feet in one minute, or, because distance fallen varied as the square of the time, 15.083 feet in *one second*.

3. *The confirmation.* The first time Newton had reasoned this out, back at home during the plague years, he had simplified things by assuming the orbit of the moon was circular, and he had used Galileo's faulty data. The value he predicted from his calculations then was close to the observed value, but still too far away to serve as a clear confirmation of the prediction. By the time he wrote the *Principia*, Newton had shown that elliptical orbits were implied by the inverse square law and he had replaced Galileo's erroneous measure of the diameter of the Earth with a more recent and more accurate figure. In the *Principia* he delights in citing Christian Huygens's measurement of how far objects at the surface of the Earth fall in one second: 15.083 Paris feet. On the basis of the assumption that the moon is affected by the apple force, Newton had made a corrected prediction that, after some twenty years, he was able to confirm.

The Principia *and Its Aftermath*

There are few works in the history of physics whose impact has been as great as that of Newton's famous work. And yet its immediate reception was not without criticism, especially concerning his notion of attractive force. But there is no doubt that the book quickly catapulted him to fame in England and abroad.

A system of the world.   If Halley had heard stories about Newton's abilities before his visit to Cambridge, when he received Newton's promised answer to his question about orbital shapes he knew there was no question about his rare talent as a natural philosopher. Most likely it was during Halley's second trip to Cambridge in the fall of 1684 that he discovered that Newton was working up an expanded treatise on the material he had sent him. With Halley's enthusiastic encouragement the work continued to grow. Halley became in effect Newton's editor, urging him on in his work, proofreading the text, and arranging all the details for printing the final product. The Royal Society, to whom Newton had agreed to send the finished product, was unable to bear the costs of printing it; consequently, Halley took that burden on himself as well.

The first volume of the work received the imprimatur of the Royal Society in July of 1686, and the whole project appeared in Latin in the summer of 1687. It was immediately recognized as a major achievement and

Isaac Newton

brought Newton great fame. Written in the format of the ancient mathematical text of Euclid, the work began with a series of definitions and axioms, the latter spiced with many corollaries. Definition 5, for example, made clear Newton's recognition of the centrality of centripetal force, the force that impels a body toward a center.

The innocent statement of the first two axioms, also called laws of motion by Newton, disguised a change in his thinking that had occurred while he was revising and expanding the treatise he had sent to Halley. Law 1 said simply: "Every body continues in its state of rest, or of uniform motion in a right line, unless it is compelled to change that state by forces impressed upon it." Here was an unequivocal claim that rest and uniform motion in a straight line were similar states, because the presence of an unbalanced force disturbed the one just as it did the other. Newton did not think that rest was *caused* by a force of any sort—it simply existed as a natural state of being. He now accepted that uniform straight-line motion was just like that—it, too, simply existed as a natural state of being. This meant that Newton had finally abandoned the idea that uniform motion was caused by an inherent active force and he now accepted Descartes's understanding of natural motion.

The way was now clear for Newton to recognize that there was more than one way for an impressed force to change the state of a moving body's natural state of being. If rest was the only natural state of being, then the only way an impressed force could change it would be to change its speed from zero to some finite amount. But if uniform rectilinear motion was also a natural state of being, force could be used to change the direction as well as the speed of the body. From this point on, a body moving in curved motion, clearly under the influence of a force, had to be regarded as accelerating just as much as a body whose linear speed was changed by the application of a force. The *Principia*'s Law II reiterated that the change of motion (or acceleration) of a mass (m) was proportional to the force that caused it, $F = ma$.

After the introductory section, Newton moved on to Book I, "The Motions of Bodies." Here he included not only the demonstration sent to Halley in November of 1684, but a total of 210 pages of propositions that unpacked all manner of detailed conclusions (many cast in general form), that arose for bodies moving under the influence of force. In Book III he came to "The System of the World." To construct a system of the world was a classically philosophical endeavor, but Newton noted that he would continue to present the material "in the mathematical way."

It must be emphasized that Newton's removal of the active force in matter that he had once thought caused uniform rectilinear motion by no means meant that he had embraced Descartes's entire mechanical approach. Newton hardly abandoned altogether the idea of an active force animating matter. As we know from his ideas on gravity and from his alchemical work, which continued even as he worked on the *Principia*, active forces were too deeply embedded in Newton's disposition to eradicate completely.

Newton knew that his readers would not permit him to make unwarranted inferences. He knew in particular that he would be hard pressed to convince them to accept the active force pulling the moon toward the Earth and the planets toward the sun. That would be difficult enough, but while drafting the *Principia*, Newton had become convinced that the force determining the planets' motions arose, as he would later put it, "from the universal nature of matter." That was equivalent to saying that not just the planets were affected by the sun's gravity, but that *every body of matter was drawn to every other body* by a force that varied as the inverse square of the distance between them. This grand generalization has become known as *universal gravitation.* Mechanical philosophers would surely not tolerate this celebration of an inherent active force of matter, because for them force was only transmitted by impact. Assigning matter the capacity to attract other matter would amount to an appeal to occult or hidden forces. Newton knew he would have to build his case carefully.

Newton had written to astronomers for data on the planets and their satellites in order to be sure that their observed positions did indeed agree with the predictions of his system. Just because he had successfully described the Earth–moon system as the result of gravitational attraction did not mean that he could simply assume it applied to all the other celestial systems. As specific data came in, he was gratified to learn that it was consistent with his claims. He showed that the paths of comets, not always elliptical but sometimes parabolic, were anticipated by his results. He showed how the oceans should be affected by the pull of the moon and the sun on the Earth, working out a rudimentary scheme explaining tidal motions. He prepared the way, in other words, for his claim late in Book III to have proven that such forces arose "from the universal nature of matter."

Newton took pains in writing the *Principia* to treat this attraction mathematically and not to assert anything about how it was caused. He hoped that he had made a case for this claim that would stand on its own. He did not want to become engaged in a debate about the mechanism by which gravitational force was transported from one body of matter to another. That, he saw, involved speculation.

The reception of Newton's attractive force. The *Principia* impressed its readers with its thoroughness, its mathematical depth, and the sheer scope of its subject matter. Prior to its appearance in England, few if any books there or elsewhere had been as widely acknowledged. But acknowledgment was not the same as agreement. Readers steeped in the tradition of Descartes and mechanical philosophy balked at the central place Newton had given to attractive force in his system. They insisted that his failure to identify a medium by which gravity's force was transmitted was equivalent to saying that it was transmitted without using any medium at all. And that, for a mechanical philosopher, was not only unacceptable—it was impossible. Christian Huygens, who called Newton's attractive force "absurd," said about him: "I esteem his understanding and subtlety highly, but I consider that they have been put to ill use in the greater part of this work, where the author studies things of little use or when he builds on the improbable principle of attraction."

If Newton's attractive force did not require a medium to be transmitted, then mechanical philosophers assumed it behaved like a psychical force. They understood that psychical powers of human beings, for example, were supposed to be transmitted from one being to another without reference to an intervening medium. Such forces were said to "act at a distance," meaning that the effect of a force exerted at one point was immediately felt at another point some distance away. Natural philosophers in the Middle Ages had not hesitated to appeal to this kind of force to explain natural phenomena. Newton's endorsement of a gravitational force that acted at a distance did not sound like a step forward.

The black year of 1693.   The sudden fame his book brought him marked a profound change from the relatively solitary life Newton had known. He met numerous prominent people, including the philosopher John Locke (1632–1704), with whom he shared his anti-trinitarian convictions and opened a discussion about alchemy. He also became enamored of a talented 25-year-old Swiss mathematician, one Nicolas Fatio de Duillier (1664–1753). For four years beginning in the summer of 1689, Newton and Fatio engaged in a protracted correspondence and even spent blocks of time with each other. The relationship, which has fueled debate about Newton's sexuality among historians, broke off abruptly in early 1693, for reasons that are not clear.

Some historians point to the breakup of the friendship as the reason for Newton's mental breakdown later that year. Extant letters to Locke show Newton to be in an extremely disturbed state, rumors of which circulated at the time. There were reports that Newton was at death's door as late as 1695. Although it is clear that Newton did endure some kind of mental trauma, there has been no agreement about its cause. While some point to the breakup with Fatio, others have suggested that Newton suffered from mercury poisoning because of repeated exposure to mercury during alchemical experimentation. Although a chemical analysis of a hair found in one of his books in the 1970s revealed a high mercury content, this explanation is not consistent with Newton's relatively rapid recovery to full vigor.

◎ Fame and Power ◎

The years around the *Principia*'s appearance were momentous for England. In 1685 King Charles II died, bringing his Roman Catholic brother to the throne. Ever since King Henry VIII had broken with Rome over the question of divorce 150 years earlier, Protestants and Catholics in England were engaged in a heated rivalry. The Anglican Church had been consolidated under Queen Elizabeth as the official Church of England in the sixteenth century, but suspicions were immediately raised any time a new monarch showed sympathy for Rome. At such times England was thrown into instability until the crisis was resolved.

When James II unwisely attempted to win freedom of worship for Catholics in England, this action united Whigs and Tories in defense of the Anglican Church. Newton, certainly no devout Protestant, nevertheless shared the widespread opposition to James's actions. In early 1687, as he finished up Book III of the *Principia,* he made known his objection to the king's position and emerged to rather sudden prominence within the university. The next year James was forced to flee England during the so-called Glorious Revolution that brought William and Mary to the throne. Newton, now a famous natural philosopher, was elected to Parliament at the beginning of 1689, where he served one year.

Mastering the Mint and the Royal Society

In Newton's correspondence from the years after the *Principia* there is more than one mention of a possible appointment to an official post. An offer was finally made in April 1696, and Newton accepted it. He was to be warden of the mint, a position that would bring him a handsome salary and would not require much work. Because the real work was usually done by the master of the mint, Newton could have continued to live in Cambridge had he so desired. He chose, however, to move to London, where he took in his seventeen-year-old niece, the daughter of his stepsister Hannah, who had recently been widowed. By aiding his stepsister, Newton also benefited himself. He grew very fond of his niece Catherine Barton, one of the very few women who affected him in any meaningful way during his entire life.

He took his new post very seriously, warming to the responsibility of apprehending and arraigning counterfeiters. He developed a reputation as a ruthless prosecutor among those he flushed out and brought to judgment. When the master of the mint died in December of 1699, Newton replaced him.

In London, Newton could easily attend meetings of the Royal Society; indeed, he was elected to its governing body twice in the waning years of the century. But Newton took no interest in the administrative concerns of the society and only rarely attended its general meetings. Then in March of 1703, Robert Hooke, a prominent presence in the society since its inception, died. Newton, who had allowed his name to be put forward as a candidate, was elected president at the age of sixty later that fall. He would not relinquish the title during the remainder of his life, becoming, as one historian has put it, "the autocrat of science" in England.

Although Newton was elected to this position of honor, he still had enemies. Hooke was gone, but Hooke's partisans were not. Those who had felt the sting of Newton's invective and those who simply resented Newton's tendency to exploit his fame in a high-handed fashion were suspicious of what he might do in a position of power. They were right to worry. As president, Newton became a dominating presence in the society, both in its administration and in its regular meetings. A number of younger men who had become devoted to Newton's natural philosophy were rewarded with professorships, and little if anything that was entered into debates about the Newtonian philosophy escaped the president's personal involvement and even supervision.

The Opticks

If Newton's rivalry with Robert Hooke had played a part in his delayed involvement in the Royal Society, it was the determining factor in the publication of Newton's larger work on light. Newton had promised himself that he would not publish this work, which had long been under preparation, until Hooke died. After the spring of 1703 the way was clear, and the *Opticks* appeared, in English, a year later. Those on the Continent who did not read English would have to wait for the Latin translation, which came out in 1706. Anyone who read it, however, found that it was easier going than the *Principia* had been. For this reason the impact of the work was as great as that of the *Principia*.

The first edition of the *Opticks* culminated in sixteen "queries," innocent-sounding questions apparently thrown out to titillate the reader's curiosity. Among other things, the queries amounted to a defense of the position he had taken in the *Principia* on attractive force. No one doubted that the questions were really Newton's answers, that when he asked: "Do not bodies act upon Light at a distance, and by their action bend its Rays?" he meant that he believed they did in fact behave in this manner. In the Latin edition that appeared on the heels of the first edition, Newton added to the queries until their total was twenty-three. He speculated on the presence in nature of a whole range of forces whose activity at a distance produced such phenomena as electricity, magnetism, and chemical interaction. His reinforcement of the necessity of active principles in nature made clear once again that he did not wish to be counted among the strict mechanical philosophers of his day.

In the last queries of the Latin edition, Newton tried to clarify where he stood in comparison to those who made up hypotheses in order to explain all things in terms of mechanical interaction. He still believed as he always had that God animated matter with powers that acted at a distance. His reference to these powers, however, made no direct appeal to mechanical interaction. That did not mean that Newton regarded the powers as mysterious qualities. They were simply phenomena that obeyed laws of nature; only their causes were hidden. But knowing the laws by which they acted was, he thought, "a very great step in Philosophy," even if he did deliberately leave their causes to be found out later. As for the first cause, Newton outdid himself: "Is not infinite Space the Sensorium of a Being incorporeal, living, and intelligent, who sees the things themselves intimately . . . and comprehends them wholly by their immediate presence to himself?"

God and Nature

Not surprisingly, Newton was criticized, especially by Continental thinkers, for what they took to be a philosophy (and theology) laced with problems. In 1710, for example, Leibniz went on record against the idea of action at a distance, which he saw to be equivalent to an embrace of miracles. He even suggested in a review of a Newtonian work that it seemed a return to "a certain fantastic scholastic philosophy," thereby hinting that Newton was trying to take natural philosophy backward rather than forward. It is understandable that Newton spoke out in the face of such accusations.

He tried once more to answer his critics about matter and attraction in a new edition of the *Principia*. In preparation for some time, it finally appeared in 1713 under the editorship of Roger Cotes, Professor of Astronomy and Experimental Philosophy in Cambridge. At the end of the work Newton waxed eloquent not only about gravity, but about its relation to God. Concerning gravity, he criticized those who insisted on speculating about its cause. He insisted on staying within the limits of his mathematical treatment, declaring in a famous phrase, "I feign no hypotheses." Concerning God, Newton declared that the beautiful system of the sun, planets, and comets could only proceed from an intelligent being who ruled over all. Newton declared that God was omnipresent not only in virtue, but also in substance.

Leibniz did not think much of Newton's abilities as a philosopher. He recognized that Newton and his editor had him in mind in their new edition of the *Principia,* and he was ready to reply. To do so he wrote to Princess Caroline of Ansbach, a young girl he had tutored at the court in Berlin who was now wife of the heir to the

English throne. Caroline had read Leibniz's *Theodicy* and had sought her former teacher's opinion of the theology of her new homeland.

Leibniz informed the princess that Newton made God into a corporeal being who used space as an organ by which to perceive things. He added that Newton also believed that God had to step in from time to time to wind up the watch of the clockwork cosmos to prevent it from running down. Newton's God "had not, it seems, sufficient foresight to make it a perpetual motion." This latter view Leibniz apparently inferred from a passage in the final query of the Latin edition of the *Opticks,* in which Newton observed that the irregular movement of comets would eventually disrupt the system of the planets "till this system wants a reformation." Indeed, Newton acknowledged that the repeated elliptical orbits of planets are never exactly identical, and he believed God occasionally caused comets to strike the sun as a means of refueling its power.

To Leibniz, Newton had demeaned God by suggesting that God's handiwork was in need of repair. Newton's God was, as one historian has characterized him, no better than a "cosmic plumber," fixing the occasional leaks that sprang in the universal system. Princess Caroline also found the notion distasteful that God had "to be always present to readjust the machine because he was not able to do it at the beginning." As she wrote to Leibniz in the beginning of 1716, she did not believe that any philosophy could give her confidence "if it showed us the imperfection of God." Leibniz's God, of course, did not need to intervene in nature, having perfectly anticipated every contingency from the beginning.

For Newton, God was intimately tied to and in complete control of the physical world, present in it by virtue of the active principles that animated matter. Newton's God could exercise infinite power and wisdom to anticipate the needs of sparrows and all other creatures, and to respond to the prayerful petitions of human subjects to intervene on their behalf in the normal course of events. The disagreement between Newton and Leibniz about how God related to nature would reverberate down through the centuries from that point on. Had God made nature perfect from the start, as Leibniz held, or was God's constant supervision of nature needed, as Newton believed?

Leibniz died in the fall of 1716, and the bitter controversies that had divided him from Newton began to subside. The next year saw a new edition of the *Opticks,* which was unaltered except for the section containing the queries. Among the eight new queries was what at first glance appeared to be a concession to his critics, the mechanical philosophers. Newton postulated the existence of "an aether, exceedingly more rare and subtile than the Air, and exceedingly more elastik and active," to explain gravity itself. But this "aether" could never satisfy his critics because it was composed of particles that repelled each other; in other words, here was action at a distance all over again.

During his last years Newton gave a great deal of attention to the study of religion, specifically the history of the ancient kingdoms portrayed in the Bible. As the end approached, he began to put his things in order. His health declining, he attended fewer meetings of the Royal Society, presiding for the last time on

March 2, 1727. As he lay dying later that same month, Newton affirmed the rebellious religious stance he had so long embraced by refusing the sacrament of the church. Three days after his death on March 20, the records of the Royal Society marked his passing with the terse announcement: "The Chair being Vacant by the Death of Sir Isaac Newton there was no Meeting this Day."

When Isaac Newton set off for Cambridge University in the early summer of 1661 there was not yet a consensus about the viability of the new Copernican view of the cosmos. Galileo had offered a reason why planets would continue to move forever around the sun in circular orbits, but Kepler had shown that the orbits were not circular. Why did the planets continue to move in elliptical orbits around the sun? Not only did Newton give the answer in the *Principia* through his laws of motion and universal gravitation, but his answer provided a means of analyzing the motions of all matter, whether in the heavens or here on Earth. Newton united natural philosophy into one comprehensive system that would dominate for the next two centuries.

Suggestions for Reading

Gale E. Christianson, *In the Presence of the Creator: Isaac Newton and His Times* (New York: Free Press, 1984).

Betty Jo Teeter Dobbs, *The Janus Faces of Genius* (Cambridge: Cambridge University Press, 2002).

Richard Westfall, *Never at Rest: A Biography of Isaac Newton* (Cambridge: Cambridge University Press, 1983).

CHAPTER 9

─────────◎─────────

Newtonianism, the Earth, and the Universe During the Eighteenth Century

The novel ideas that came into science in the seventeenth century were incompatible in many ways with the more comfortable cosmos of former times. Copernicus had already moved the Earth off to the side, away from the center of the system of spheres that had always provided humans a home. But even in the early versions of the Copernican system, including that defended by Galileo, the cosmos at least remained finite in extent. After Descartes and especially after Newton, it was no longer possible to insist that space did not extend infinitely in all directions.

It took some time to build a consensus about the meaning of Newton's achievement. After all, there was much more about it to disagree with than just the question of whether the universe had a center or not. The linchpin on which all depended in Newton's system was his notion of an attractive force that acted at a distance. For mechanical philosophers who continued in the heritage of Descartes, this was a major stumbling block. The transmission of Newton's force appeared to make use of an intervening medium that was occult, and that was simply unacceptable to them.

◎ The Rise of Newtonianism ◎

In spite of the fame Newton enjoyed among his fellow British citizens, his system initially found few followers abroad. After 1730 Newton's system began to attract followers—particularly in France—who defended a worldview that has been called Newtonianism. But prior to 1730, the continuing influence of René Descartes in France and Gottfried Leibniz in the German states was sufficient to assure that the Cartesian and Leibnizian worldviews provided strong competition for Newton's thought. Only in Holland did Newton's system find avid defenders across the English Channel.

Competing Systems of Natural Philosophy

The basic assumptions individual natural philosophers made about how nature worked determined differences among them that carried implications throughout their systems. In the early eighteenth century these differences produced an important debate about force and action in nature and also involved the issue of God's relationship to the natural world.

Cartesians, Leibnizians, Newtonians. The most important issue separating Newton's system from those of Descartes and Leibniz remained an understanding of the nature of force. The Cartesian position was clear: force was a push or pull that acted on material objects by means of material contact. All natural effects were due to mechanical motions of matter, making the appeal of the Cartesian philosophy its intuitive clarity. Cartesians in the first half of the eighteenth century did not feel obligated to accept the specific mechanical motions Descartes had used to explain individual phenomena, but they did not doubt that things like magnetism resulted from *some* combination of such motions. Descartes exerted a strong hold on the French mind because his readers understood him to have clarified the basis for intelligibility itself in physics.

Because followers of Descartes insisted that force was transmitted only through collisions of matter, they refused to associate force with nonmaterial agencies in nature. Nature was a realm of the material. It was unacceptable to offer explanations of natural phenomena that depended on spiritual or nonmaterial occult agencies. To Cartesians, the assertion that force acted at a distance was equivalent to an appeal to a nonmaterial agency.

Leibniz's system was represented after his death by the natural philosopher Christian von Wolff, whose work was available to German readers within a year of Leibniz's death in 1715. Wolff regarded Descartes's explanations of the physical world as helpful but limited. The Cartesian approach applied to what we see, but it did not relate to the deeper reality Wolff believed lay beneath appearances. For the superficial level of appearances, Wolff was content to embrace Descartes's mechanical interactions to make the appearances intelligible. To explain how force acted, he too rejected the occult agencies of medieval Scholastic thought (and therefore also action at a distance) in favor of forces transmitted only through contact between masses. Like his mentor Leibniz, he described the physical world as a clock designed by God to work perfectly. Wolff emphasized Leibniz's appeal to the principle of sufficient reason, according to which we understand the existence of something when we find the reason for it. Because everything has a sufficient reason, we can use our reason to show that the world has been made perfectly.

The difference between the Leibnizians and the Cartesians emerged at the deeper level of reality's basic components. Leibniz had held that matter was not equivalent to extended space, as Descartes taught, but was made up of unextended points he called monads. Monads were nonmaterial metaphysical entities that resembled souls. They were the *source* of the force; indeed, they were the source of all the activity that accompanied matter. By making a distinction between the source of force and the means by which it was transmitted, Leibnizians were both critical of Newton's action at a distance and of Descartes's banishing of spirit from the natural world.

Newton's cause was taken up abroad by the Dutchman W. J. 'sGravesande (1688–1742), who published an introduction to the philosophy of Newton in 1720. 'sGravesande answered those critics of Newton who asserted that his action at a distance amounted to a return to occult causes by declaring that gravity was not the cause of anything. It was an effect. Gravity was the name we give to the movement of bodies toward one another when left to themselves. According to 'sGravesande, physics should focus its attention on the results of experiments rather than try to devise grand causal explanations. Like other Newtonians in England, he ignored Newton's own attempt to find a cause for gravitational force in the special kind of ethereal substance that Newton made public in the 1717 edition of the *Opticks*. When Newton died in 1727, many who defended his system shunned the question of the cause of gravity, understanding the system to rest simply on Newton's laws of motion and a gravitational force that acted at a distance according to the inverse square law.

The *vis viva* controversy. Over the course of the eighteenth century Cartesians, Leibnizians, and Newtonians became embroiled in a disagreement known as the *vis viva* controversy. It centered on the question of whether force in the universe could be lost—that is, whether the total amount of force in the cosmos could become diminished, in which case the cosmos, left to itself, would run down and eventually come to a standstill. To many this prospect was inconsistent with their understanding of God's creative abilities. But if force was conserved and could not be lost, how was force to be measured?

In his *Principles of Philosophy*, Descartes had asserted that because God was unchangeable, he "conserves the world in the same action with which he created it." Descartes envisioned the universe as he imagined God saw it from the outside—a realm filled with material objects in motion. All this motion, which involved many collisions of matter, constituted the world's action or activity. Descartes felt that God had invested this activity in the world at the Creation and that he held it constant. The constant exchange of motion over time among portions of matter constituted the history of nature itself. Descartes felt that the universe was a machine that would not run down because, in spite of the exchanges, God made sure that no motion was lost. The sum total of all the activity always remained the same because God had given to individual motions of matter the property, as Descartes put it, "of passing from one to the other, according to their different encounters."

But how was one to measure this "action"? Discussion of this problem continued into the eighteenth century and beyond, pitting those who preferred Descartes's measure—something he called the "quantity of motion"—against others who opted for something Leibniz called *vis viva*.

When Descartes asked himself what might be a measure of the quantity of motion, he thought about the force that a piece of matter exerted when it encountered another piece of matter. That, he reasoned, obviously depended on two factors: how big the mass was and how fast it was moving. He concluded that the force of motion could be expressed as the product of mass (m) and velocity (v), and he determined that this was a measure of the quantity of motion.

If we consider just two pieces of matter moving toward each other, we can calculate the force of motion of the first piece ($m_1 v_1$) and also that of the second ($m_2 v_2$). Adding these two amounts, we have the total force of motion of the two ($m_1 v_1 + m_2 v_2$). Descartes held that, after the collision, this total amount remained the same, although the individual velocities of the two pieces of matter might change. Whatever velocity was given up by one piece of matter was given to the other, so that the total sum of the masses times their velocities remained the same. What happened in the case of just two pieces of matter also happened in every other collision in the universe. The sum total of what Descartes identified as the force of motion remained the same, while the changes in the velocities of individual pieces of matter due to collisions constituted the activity of the universe. For Descartes, God's immutability meant that the total force of motion in the universe was conserved and the universe would run forever.

There are problems with Descartes's claim, the most obvious of which is that it does not work for what are called inelastic collisions. If two equal blobs of clay move directly toward each other at equal velocities, when they collide they do not rebound at the same velocity but stick together, and the motion stops. What happened to the total force of motion in this case? It would appear that it has not been conserved but destroyed. To incorporate situations like this into Descartes's analysis, his follower, Christian Huygens, asserted that it was necessary to specify the direction in which the masses were moving. In other words, the forces of motion of masses moving directly toward each other must be considered opposite in sign.

If in the above case the masses are equal, and if the first mass of clay is assigned a positive force of motion ($+ mv$), then the second would have an equal negative force of motion ($- mv$). Adding up the total before and after the collision would give zero in both cases. Had the masses not been clay, but some perfectly elastic substance, then the total force of motion would still have been zero before and after the collision, except that the velocities of the two equal masses would have changed sign as they rebounded from each other. The contrary motions God had put into the universe at its beginning balanced each other in the end.

This improvement made by Huygens, however, still left a major problem. With every inelastic collision there would be less motion in the universe. That would mean that the actual motion in the universe was running down, a result unacceptable to Descartes. He had proposed his idea in order to guarantee that the machinery of the universe would continue to run.

In an article in 1686 entitled "A Brief Demonstration of a Notable Error in Descartes," Leibniz pointed out that Descartes's measure of the force of motion would not, in fact, prevent the universe from running down. He proposed a different measure of the force of motion, something he called *vis viva*, or "living force."

The problem of the running down of the universe was an issue as long as the force of motion was regarded as a signed quantity—one that could be positive or negative. As such one force could destroy another, diminishing the total God had originally invested in his creation. Leibniz proposed that a better measure of the force of motion was proportional to the mass times the square of the velocity (mv^2), which was always a positive quantity. He claimed that in inelastic collisions, like the one involving blobs of clay, the *vis viva* was not destroyed when the pieces of clay stopped moving after collision; rather, the motion was transferred to the particles that made up the clay. So the total motion continued at another level, the amount of *vis viva* in the universe remained the same, and the universe did not run down.

Many people did not accept *vis viva,* if for no other reason than it was too abstract a notion. Nor was Leibniz persuasive with his explanation of why *vis viva* was not lost during inelastic collisions. Among those who came out in favor of *vis viva* as a measure of the force of motion was the Dutch Newtonian, 'sGravesande. His major contribution to the discussion—which resulted from his wish to base conclusions as much as possible on experiments—was to think of measuring the *effect* a moving mass might have, rather than merely the force it might exert. Thinking of the force of a mass in motion as the *effect* (or damage) the mass produced in a collision, as opposed to the *push* exerted during a collision, proved to make a difference.

'sGravesande did a series of experiments in which he dropped masses onto clay and then measured the dents that were made. He varied the heights and the weights of the masses, measuring the various dents produced. He found that the impressions in the clay were the same if, when he used a mass with half the weight of another (although of the same size and shape), he dropped it from twice the height. Descartes, of course, would conclude that, if the dents were the same in the two cases, then the measure of the motion should be mv in both cases. But 'sGravesande showed that because the lesser mass was dropped from a greater height, it hit the clay at a greater speed ($\sqrt{2}v$). He demonstrated that the product of the lesser mass ($\frac{1}{2}m$) and the greater velocity ($\sqrt{2}v$) was not mv, but $\frac{\sqrt{2}}{2}mv$. He concluded therefore that Descartes's measure of the force of motion, mass times velocity, had to be in error.

'sGravesande said the correct measure of the force of motion was $\frac{1}{2}mv^2$. This measure covered both of the preceding cases. In the first case, when the mass was m and the velocity was v, his formula gave $\frac{1}{2}mv^2$. In the second case, when the mass was $\frac{1}{2}m$ and the velocity was $\sqrt{2}v$, the product of one-half the mass times the velocity squared also gave $\frac{1}{2}mv^2$. So for 'sGravesande the measure of the force of motion causing the dent made in the first case *did* equal that causing the dent made in the second. 'sGravesande had shown to his own satisfaction that Leibniz's *vis viva* was a better measure of the force of motion than Descartes's quantity of motion. By siding with Leibniz here, 'sGravesande not only opposed the Cartesians. In this instance he also went against Newton, who did not regard *vis viva* as anything real.

Just as Leibniz had found, 'sGravesande discovered that not everyone immediately agreed with him; in fact, the debate about conservation of quantity of motion and of *vis viva* continued throughout the eighteenth century. The discussion, recall, had been initiated in a theological context having to do with God's preservation of action in the world. With the exception of those who agreed with Newton (who felt that God would step in to correct the universe if it ran down sufficiently), it was important to the participants in the discussion to find an explanation that would prevent the universe from slowly degrading.

The Growth of Newton's Reputation

Newton's work finally came to the attention of a wider public in France through a popular account of its basic conclusions that appeared in 1733. In addition, a number of issues came to the surface after 1730 whose outcomes promoted Newton's reputation as a man ahead of his time. Not only had he seen farther than those who came before him, but in some cases he appeared to have anticipated solutions to problems that arose only after he had departed. The accumulated effect of these developments contributed to the emergence by the latter part of the century of a prominent group of French Newtonian natural philosophers.

Voltaire

The popularization of Newton in France. The introduction of Newton's ideas to many in France occurred in 1733, when François Marie Arouet, who had taken the pen name Voltaire, published his *Philosophical Letters.* The French authorities had imprisoned Voltaire earlier in his career because of satirical things he had written about the French government. When he insulted a powerful nobleman in 1726, he was given the choice of another stint in prison or a period in exile. He chose the latter, living in England for the next three years.

While in England he studied the customs of the English, their form of government, and the ideas of their philosophers, and he was especially drawn to the work of Newton. In his *Letters* he praised Newton, whom he called "this destroyer of the Cartesian system," and went on to present a comparison of Newton's system to that of his countryman Descartes. After declaring that Newton had proven by experiments that Descartes was wrong about the universe being filled everywhere with matter, Voltaire proclaimed that Newton "brings back the vacuum, which Aristotle and Descartes had banished from the world."

In the *Letters* Voltaire carefully explained the role played by gravitational force, "the great spring by which all Nature is moved," reproducing a summary of Newton's proof that the moon and planets are held in their orbit by this force. The reader quickly realized that in Voltaire's view Newton's system was far superior to any other. Along with his preference for Newton, it was clear that Voltaire also preferred English customs, laws, and society to their French counterparts. The message of the book got Voltaire into trouble once more and he again had to leave Paris. He took refuge in the independent province of Lorraine at the chateau of the marquise du Châtelet, a friend he had recently met and with whom he collaborated until her death in 1749.

Du Châtelet also contributed to the popularization of Newton's thought. She collaborated with Voltaire on the publication in 1738 of *Elements of the Philosophy of Newton,* a more complete exposition of the Newtonian system than Voltaire's earlier work. She also completed a translation into French of Newton's *Principia,* which appeared after her death. Although du Châtelet aligned herself with the Newtonian "party," as the growing number of French defenders of Newton called themselves, she was not a slavish follower of Newton. Persuaded of the merits of Leibnizian metaphysics, she also published a book on Leibniz's system in 1740.

The controversy over the shape of the Earth. The question about whether the Earth's shape resembled an egg or a pear provoked a controversy that ended up pitting the systems of Newton and Descartes against each other during the 1740s. Earlier in the century a long-standing project to map the kingdom of France had uncovered discrepancies in the terrestrial lengths of degrees of longitude. The leader of the project repeated the measurements to confirm their accuracy, and announced to the Paris Academy in 1718 that the results meant that the Earth was not spherical, as everyone had assumed, but that it had a slightly elongated shape, something like that of an egg. The problem was that both Newton and the Cartesian Christian Huygens, basing their conclusions on calculations made from their respective systems of mechanics, had much earlier predicted that the Earth should bulge slightly at the equator, like a pear. Initially, the question had been whether to believe the predictions of natural philosophers or a claim that appeared to rest on careful measurement. In the 1730s and 1740s the question became something very different.

Pierre-Louis Moreau de Maupertuis (1698–1759), son of a recently ennobled merchant, went to Paris as a teenager to study philosophy, but soon found his real interest was mathematics. In the early 1720s, while Maupertuis was learning the intricacies of higher mathematics, he became aware that the disagreement about the shape of the Earth bore on theoretical systems he had been studying. In the summer of 1728 he spent twelve weeks in London, where he immersed himself in Newton's work.

Between 1732 and 1736 Maupertuis turned his attention to the question of the shape of the Earth. When it was decided that new expeditions would be sent out to determine once and for all whether the Earth was flattened or elongated at the poles, Maupertuis led the group that traveled to the proximity of the North Pole to make measurements of the distance along a longitudinal line between two established latitudes. Another group went to the equatorial region of Peru to do the same thing. Because the distance between the equator and the North Pole is divided into equal degrees of latitude, any variation in the measured distance between two sets of latitudinal positions would indicate that the Earth was not spherical. A distance less than expected would indicate an elongation, and greater than expected would mean a flattening.

There was no reason why the issue of the Earth's shape had to be regarded as a disagreement between the systems of Descartes and Newton; there were partisans on both sides that made arguments for an Earth flattened at the poles. Nor did most of those arguing about the Earth's shape initially regard it as a test between these two rival systems. But Maupertuis had taken on the role of a crusader for the Newtonian system, on the one hand recruiting younger impressionable academicians to his side, and on the other ridiculing the Cartesians in the Paris Academy.

It was a decade before the expedition to Peru returned with its findings. By then the group that had gone to Lapland with Maupertuis had been back for eight years with measurements that proved consistent with an Earth flattened at the poles. Maupertuis exploited these results to the advantage of his defense of Newton; indeed, his duplicity in the affair earned him numerous enemies, some of whom criticized him for defending a foreign English system over the French system of Descartes. Voltaire also helped polarize the issue by portraying Maupertuis as France's Galileo in his struggle to promote Newton's system in Paris. The upshot of the matter was that the Newtonian party used the incident to make Newton's system appear superior to that of Descartes.

Questioning the inverse square law.   Newton's reputation in France also grew because his system survived direct challenges. One such challenge surfaced in the late 1740s when a brilliant young French mathematician announced that Newton's famous inverse square law was incorrect. Alexis Claude Clairaut (1713–1765), son of a Parisian mathematics teacher, was among the gifted young mathematicians who joined the Newtonian party of Voltaire and Maupertuis. Clairaut's challenge to Newton grew out of his interest in another theoretical challenge—what has become known as the three-body problem.

Newton's original consideration of the problem of the moon's motion had considered the gravitational interaction of the Earth and the moon. But if the more general conclusion he had come to in the course of solving that problem was true—namely, that all matter attracts all other matter according to an inverse square law—then an object like the moon would be affected by a large body like the sun at the same time it was being attracted by its much nearer neighbor, the Earth. In fact, there were irregularities in the moon's motion that Clairaut and others hoped to explain by taking the *three* mutually attracting bodies into account. The competition for the first successful solution to this three-body problem extended beyond France into Switzerland.

Clairaut and his competitors were able to write the equations that described the interacting inverse square attractions of the three bodies, but they then determined that solving them directly was impossible. The best Clairaut or anyone else could do was to devise a means of approximating a solution. The differing methods of approximating a solution used by those engaged in the problem all showed the same unexpected result: certain positions the moon was predicted to have based on inverse square attractions of the three bodies were way off from the positions the moon was observed to have. Clairaut concluded that the discrepancy was due to the inverse square law itself—it must not be correct as Newton had stated it ($f \propto 1/r^2$); it should rather have the form $f \propto (1/r^2 + 1/r^4)$. (Because r^4 is a huge number, the correcting factor of $1/r^4$ would be tiny.) His public challenge to the authority of the great Isaac Newton, made in an announcement to the French Academy on November 15, 1747, stirred a pot already boiling about the merits and demerits of the Newtonian system.

It was not long before Clairaut realized that he had made an error in approximating a solution to the Earth-moon-sun problem. In the course of working through the complicated equations he had made a mathematical simplification that seemed harmless, but on closer inspection it proved to make a big difference in the outcome. Clairaut realized that his competitors had also employed the same simplifying assumption and that this explained why they had all come to the same result. When he corrected his mistake, his approximation of how three interacting inverse square forces affected the moon produced results that agreed with the observed positions of the moon. Less than a year and a half after his first dramatic announcement to the French Academy he made a second one: there was no need to correct Newton's inverse square law after all. To Newton's supporters it seemed that the master had known better all along.

Halley's Comet.   Newton's reputation received yet another boost during the 1750s from a prediction that had been made a half century earlier by Edmund Halley. Halley had concluded that a comet that had appeared in 1682 had a path similar to comets that had been observed on four previous occasions, the earliest from 1456. He had determined from the similarity of the paths and from the time between appearances that all the observations had been of the same comet and that it would reappear next about 1758. He refused to be precise about the exact date of the next return of the comet because he could not be sure how large planets such as Jupiter and Saturn might affect the path the comet traversed.

Because Clairaut had become the leading expert on the three-body problem, he decided to apply his knowledge of Newtonian mechanics to the problem that Halley could only anticipate. With assistance from a few colleagues, he calculated that the comet would be at its closest point to the sun within thirty days of April 15, 1759. He announced this prediction in the fall of 1758 to the Paris Academy. The comet actually arrived at its closest point two days outside of this thirty-day margin of error, on March 13, but no one considered this an error on Clairaut's account. The appearance of a comet was something even people on the street were interested in. To them, Clairaut's prediction meant that he must understand the motions of heavens better than anyone before him had. Being hailed as a new Newton in the public press now brought undisputed fame to the Frenchman, Clairaut.

Perfecting celestial mechanics.   Another irregularity in the moon's motion led to the solidification of support for the Newtonian system among French natural philosophers. The problem also originated in Halley's study of past astronomical events, this time in a paper he had read to the Royal Society in 1693 about the dates of ancient eclipses of the moon. If the machinery of the heavens that Newton had described were run backward, then the position of the moon would not have resulted in an eclipse at those times when eclipses had been recorded. Halley suggested that one way to reconcile the discrepancy was to assume that the moon's orbit had shrunk slightly in the intervening time. As the orbit shrank the moon would revolve at a faster pace around the Earth, thereby decreasing the length of time in a month and providing a means to eliminate the discrepancy mentioned.

The gradual shrinkage in the moon's orbit was in fact confirmed by subsequent observations. While this strengthened Halley's suggested solution to the problem of ancient eclipses, it created another problem: it implied that the Earth-moon system was unstable, that it was running down. Yet another brilliant young French Newtonian took up this problem, this time from the generation after Clairaut. Just before the outbreak of the French Revolution, 38-year-old Pierre Simon Laplace (1749–1827) showed that the planets in the solar system exert a gravitational effect that alters the shape of the Earth's orbit very slightly. As a result of the Earth's changing position with respect to the sun, a corresponding variation in the sun's effect on the moon is introduced. The presence of this set of perturbing factors is so small that it could only be detected over long periods of time.

But did that mean that Laplace had shown the system to be unstable? In fact the opposite was the case. Laplace demonstrated that after many years the factors that caused the moon's orbit to shrink would begin to move it in the opposite direction. The moon's orbit underwent an oscillation in and out, one cycle of which took thousands of years. Newton's universe was stable after all. It would not run down. From this and his other work on Newtonian celestial mechanics, Laplace imagined a determined system of the world, of which we would have perfect knowledge if we could but know "the state of all phenomena at a given instant, the laws to which matter is subjected, and their consequences at the end of any given time." Others might think it presumptuous and arrogant to speak with such boldness. To Laplace it was simply the perfecting of what Newton had begun.

◎ An Expanding Cosmos ◎

As we learned in Chapter 2, the notion that God may have created other worlds was long a theological concern. This historical reference to other worlds referred specifically to heavenly homes for living beings. With the growing consensus among natural philosophers of the seventeenth century that Copernicus had corrected an ancient error about the structure of the cosmos, interest in this specific theological issue increased. Natural philosophers weighed in on the implications for theology of life elsewhere.

Other Worlds

Over the course of the eighteenth century speculation about life elsewhere in the universe continued unabated. There were those who objected to the general enthusiasm in favor of pluralism, as historians have dubbed belief in the existence of other inhabited worlds. The philosopher David Hume based his skepticism on rational analysis. But most of those who objected did so mainly for theological reasons, the primary one of which centered on the meaning of Christ's death on the cross. If there were millions of worlds similar to the one in which we live, then why would God give his son for this one?

By far, however, writers in the eighteenth century—whether concerned with theological or philosophical issues—readily accepted the existence of life beyond Earth. They established a consensus of opinion that blurred the lines that otherwise divided religiously minded individuals from each other or that separated religious thinkers from more secular thinkers. It was a consensus that would enjoy remarkable staying power.

The rise of the new science had an impact on theological practice beyond merely stimulating interest in the specific question of extraterrestrial life. It also raised the status of natural theology, a branch of theology that had not previously enjoyed prominence. Reasoning in natural theology did not commence with the truths of a

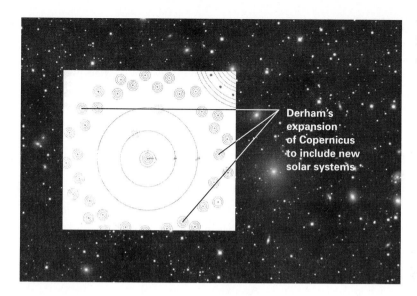

divine revelation, as in other branches of theology. The natural theologian began instead with the knowledge of nature. On the basis of evidence of intentional design uncovered in the natural world, natural theologians reasoned to the existence of a divine designer. The development of works of natural theology was not confined to individuals formally trained in theology. Many natural philosophers were delighted that their pursuit of the new science was not only compatible with their personal belief in God, but provided impressive new forms of the argument from design—that it was possible to infer the existence of a designer from evidence of design in nature.

Among the early works of natural theology devoted to the plurality of worlds was William Derham's *Astro-Theology*. Derham was an Anglican clergyman who became chaplain to the Prince of Wales in 1715, the year his *Astro-Theology* appeared. He wrote the book as a companion to his *Physico-Theology* of 1713, in which his natural theological argument centered on knowledge of terrestrial nature. Both of his works were enormously popular in Britain throughout the eighteenth century, running to fourteen English and six German editions by 1777. In the *Astro-Theology* he used Newton's work on astronomy not only to justify a Copernican arrangement of the planets around our sun, but also to infer what he called a new system that went beyond the Copernican. Derham argued that "there are many other Systemes of Suns and Planets, besides that in which we have our residence; namely, that every Fixt Star is a sun and encompassed with a Systeme of Planets." What made Derham confident that so many solar systems existed was that they were "worthy of an infinite Creator."

Nebular Hypotheses

Observers of the heavens, even before the invention of the telescope, had identified more than just stars and planets. Some of the lights in the heavens appeared as luminous cloudy patches, which could be mistaken for comets until the observer noticed that the patches retained their position just as stars did. Ptolemy recorded seven such objects, Tycho Brahe six. Included in the 1690 star catalog of the Polish observer Johannes Hevelius were sixteen "nebulous stars." Others compiled lists of such nebulae prior to 1750, but it was not until the second half of the century that they became the focus of attention as possible island universes of their own.

In speculations from the time, the concept of nebulae had two distinct meanings. Writers referred to nebulae as the distant luminous patches in the sky that, if seen from the proximity we have to our own Milky Way, appear as a cloudy film. Nebulae in this sense denoted systems of stars. Another meaning concerned the matter of the primitive universe, understood as having been distributed in a thin homogeneous fluid. Although different from each other, both meanings contributed to the speculations that ultimately became known as nebular hypotheses.

Kant's natural history of the heavens.   Among the most famous names in the history of Western philosophy, Immanuel Kant's ranks at or near the very top. His philosophical contributions, which are treated in Chapter 14, derive from work done primarily during the last half of his adult life. As a young man he was interested in natural science. He studied the systems of both Wolff and Newton at the university in Königsberg, where he later taught. As a young man of thirty-one, he wrote the *Universal Natural History and Theory of the Heavens,* published in 1755. It contained his ideas about cosmology, the theory of the heavens, and cosmogony, an account of the origin and development of the universe. In both enterprises Kant used Newton's principles, which he invoked in the subtitle of his work, as a starting point for his reasoning. But Kant went beyond what Newton was willing to consider. He suggested that there were systems of stars (systems of systems) that rotated around a central mass. Our Milky Way, he contended, was one such massive system whose orbital motion prevents it from collapsing into its center. According to Kant, nebulae were other "Milky Ways," each presumably containing systems of stars with their planets.

The cosmogony in the second part of the work utilized natural laws more than it did a Creator to explain how such massive systems had originated. Nebulous matter diffusely distributed in an infinite universe gradually coalesced into dense masses as a result of gravitational force, first into rings and then into spheres. Kant thought he could account for the rotational motion that commenced around the primitive centers of gravitational force by introducing repulsive forces acting between particles of mass. This same process occurred at various levels, first in planets that ended up revolving around central suns, next in suns rotating around their own massive centers, and finally in other "Milky Ways" rotating around the center of the universe itself. Although he did not explain how an infinite universe could have a center, he imagined that the formation of "worlds without number and without end" had begun near this center and spread out as the order introduced by gravitational force gradually conquered the chaos of diffuse matter. He referred to the "mountains of millions of centuries" that it would take for the formation of worlds to reach perfection. While he conceded that God was the ultimate explanation for the existence of order that conquered chaos, his interest was primarily in explaining how that order worked.

Laplace's nebular hypothesis and the solar system.   Another cosmogony of the second half of the eighteenth century appeared in 1796. It was written by Pierre Laplace barely a decade after his dramatic demonstration that the solar system was

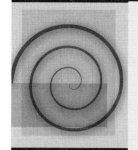

Was Newton Right About Hypotheses?

The NATURE of SCIENCE

Newton's famous claim, "I feign no hypotheses" (see Chapter 8), served as a justification for many in the late-eighteenth century to refuse to engage in speculation, as if natural philosophy should content itself only with facts. The prize competitions of numerous scientific societies from this time, for example, specified that acceptable answers should confine themselves to matters of fact alone.

Natural science relies on careful empirical observation and experimentation, but it often also requires creative imagination to formulate theories that in turn suggest new ideas. As is evident in this section, even during a time when speculation was widely frowned on, leading natural philosophers willingly constructed hypotheses to explain how the cosmos had come about. Newton's claim may have been invoked by some in this time to justify their anti-hypothetical viewpoints, but in so doing they took his public statement out of context. As we have seen, Newton did speculate about numerous matters on many occasions.

not running down. Laplace restricted his attention to the origin of the solar system, rather than the entire universe, but its more limited context did nothing to diminish the grandiose implications of his conclusions.

By the time Laplace had formulated his hypothesis the astronomer William Herschel had already used powerful new telescopes to confirm the presence of hundreds of new nebulae in the heavens. They appeared as structureless masses of a finely distributed substance, some of which displayed apparent condensation in the center. Laplace inferred that they represented various stages in the formation of planetary systems and reasoned that our own solar system had originated from a similar process. In his book, entitled *An Exposition of the System of the World*, he postulated a rotating, primitive nebulous solar fluid that contracted as a result of gravitational force into a central mass surrounded by smaller revolving masses.

Laplace argued that the origin of our solar system from a rotating nebula explained why all the planets and known satellites revolved around the sun in the same direction and why they were all in planes only slightly inclined to each other. Newton had used this same coincidence of circumstances to argue that the solar system had originated from the direct intention of God. Laplace, too, believed in God, but he did not believe that God's presence was required to supervise the workings of the cosmos. According to an account that originated in the middle of the nineteenth century, when Napoleon asked him where God appeared in his system of the world, Laplace replied: "Sire, I have no need of that hypothesis."

◎ The Earth as a Cosmic Body ◎

It has been said that some of the natural philosophers of the eighteenth century wanted Newton's physics without Newton's God. More and more they preferred Leibniz's God, who created laws that made the universe run by itself, to Newton's repairman God. In 1692 Richard Bentley defended his friend Newton's position against what he called *deism,* a term that has come to characterize the belief that God was necessary to create the physical and moral world orders, but not to superintend them. The embrace of both Newton's physics and deism in the second half of the eighteenth century by many French natural philosophers confirmed that it was indeed possible to separate Newton's description of how the heavens worked from religious views about their origin and governance.

Causal Theories of the Earth

Buffeted about by the impersonal forces of gravitational attraction, the Earth at the midpoint of the eighteenth century no longer enjoyed anything close to the exalted position it had once held in God's Creation. The Earth was not exempt from the laws astronomers used to understand heavenly bodies. It was inevitable that the same attitude that led some natural philosophers to explain naturalistically how the heavens originated would lead others to consider how the Earth came to be as a result of the operations of natural laws.

To the average person the Bible was clear about how long God's creative activity had taken—seven days. Furthermore, the Bible provided clues about the duration of history. A seventeenth-century English scholar, Archbishop James Ussher, had used his extensive biblical knowledge of the genealogies of Adam's lineage to determine the date of Adam's creation. Ussher's calculation of 4004 B.C. as the date of the creation of the universe confirmed the impression of most that the universe and the Earth were approximately six thousand years old.

But as early as the seventeenth century natural philosophers had supplemented the biblical record with what became known as theories of the Earth—conjectures about the *causal means* God might have used. In the earliest of these theories of the Earth the physical causes and mechanisms identified did not challenge the Genesis chronology; on the contrary, the older theories of the Earth used the new science to support it. That would change in the eighteenth century.

The *Telliamed* of Benoît de Maillet.   Sometime between 1692 and 1718 France's ambassador to Egypt, Benoît de Maillet (1656–1738), composed what he called "a new system on the diminution of the waters of the sea." It was an attempt to explain what had caused the Earth and life to develop as they had. By the time the manuscript reached the public in 1748, de Maillet had been dead for ten years. He had certainly intended to see the work published during his lifetime, but unfortunate delays had frustrated his plan.

De Maillet had traveled widely in the Mediterranean, taking with him an intense curiosity about the features of the Earth's surface in the regions he visited. He did not limit his interest in the Mediterranean area to its geography. He mastered Arabic, read the histories of Arabic writers, and became familiar with historical landmarks. His travels exposed him to cultures whose understanding of history, including the history of the Earth, was very different from that of Christian France. He determined that he would incorporate this broader perspective into his new system.

De Maillet's grandfather had become convinced that the sea was diminishing, through observations he had made of the seashore near the family home. From his grandfather's observations and from knowledge he acquired on his travels, he composed

Telliamed (which was de Maillet's name spelled backward). De Maillet knew his manuscript would test the limits of acceptability, so he tried at the outset to deflect criticism by attributing the views expressed in the book to a pagan foreigner. The subtitle declared that the work consisted of conversations between a French missionary and a philosopher from India named Telliamed about the formation of the Earth being due to the diminution of the sea. De Maillet utilized mechanical interactions of vortices together with his own observations to create a specifically non-Christian cosmogony.

From the alleged Indian understanding of the Earth's past the French missionary learned that the Earth was originally covered with water, whose currents carved out the mountains beneath its surface. The depths of the primitive seas gradually decreased, exposing the highest mountains, which immediately began to wear away. Eroded material from the shore settled onto the ocean floor as sedimentary rock, which, as the sea level continued to drop, was exposed as new mountains. As the process of diminution continued, more dry land emerged and on it, life began.

Telliamed himself did not invoke the direct act of God to explain the successive appearances of animal life. He did not give details, but he maintained that various forms of aquatic animals had changed during the time the sea was gradually receding in accordance with natural process. Air above the seashore was so moist "that it must be considered an almost equal mixture of air and water," breathable, for example, by flying fish. While escaping a predator or having been thrown onto the land by the waves, such fish found their features altered. "The little wings which they had under their belly, and which like their fins helped them to walk on the sea bottom, became feet and served them to walk on land."

Clearly such processes had taken a great deal of time, far longer than six thousand years. Telliamed in fact estimated from structures in ancient Carthage that the rate the sea level had dropped from earlier times to his day was three feet every thousand years. Using this rate he concluded that over two billion years had passed since the primitive waters had begun to recede. Humans themselves were over 500,000 years old!

The public immediately saw through De Maillet's tactic of camouflaging his ideas in a pagan philosophy. The reaction was outrage. But retribution could not be exacted on de Maillet, who was long dead. Another writer who speculated publicly that same year had no such refuge.

From *Natural History* to *Epochs of Creation*.   Although Georges-Louis Leclerc (1707–1788) was born into aristocracy, the name by which he has become known to history—Buffon—was given him by King Louis XV of France in 1773 for what he had achieved during his lifetime. The king bestowed on him the title Comte de Buffon (based on the name of an estate he inherited from his mother when he was twenty-five), because of his extensive accomplishments as a natural philosopher. Buffon is remembered for his monumental *Natural History,* a multivolume work about the living world that even nonspecialists could understand. The first three volumes were published in 1749.

In the first volume Buffon included a history of the Earth because, he said, it was "the history of nature in its most ample extent." It became immediately clear that Buffon believed the Earth was extremely old; he announced that the more recent changes of the previous few thousand years were insignificant compared with those that occurred in the ages following the Creation. Because Buffon viewed the Earth as just one of the planets, its history was tied to that of our solar system.

Readers of the first volume of Buffon's *Natural History* encountered an intriguing idea of how the Earth originated. Buffon explained what caused all of the planets to circle the sun in the same direction and within the same plane: a comet had struck the sun at an oblique angle, knocking off huge pieces of mass that settled into orbits at various distances away and then began to cool. To justify his idea that a comet might have struck the sun he invoked the authority of "the great Newton," who had suggested that comets occasionally collided into the sun and refueled the sun's power. Buffon concluded that as it cooled, the Earth became covered with water from the condensing atmosphere. Tidal motions of this primitive universal sea carved out mountains and valleys beneath the surface. He was reluctant to say how dry land came about as time passed, although he did make reference to likely violent revolutions.

Some reviewers of Buffon's work immediately noted that his explanation of the origin of the Earth was incompatible with that given in the book of Genesis. In the biblical account the Earth was created *before* the sun, moon, and stars, which did not appear until the fourth day of creation. One reviewer, who regarded Buffon's work as outlandish heresy, noted that it required a world "far older than Moses made it out to be." Buffon had not said exactly how old, but the reviewer asked whether there really was any difference between Buffon's view and that of authors who believed the world was eternal.

Early in January of 1751 the faculty of theology at the Sorbonne in Paris notified Buffon of some fourteen propositions from the first three volumes of the *Natural History* that they regarded as reprehensible. Among these were Buffon's speculations on the Earth's fiery origin. In order to avoid censure, Buffon published a retraction in the fourth volume of his *Natural History*, which came out in 1753.

It is clear from his continuing work that Buffon's recantation was not genuine. He knew about recently published measurements of the heat in mines and hot springs that had been made in order to support the claim that the center of the Earth was hot. He also knew of Newton's estimation that an iron sphere the size of the Earth would take more than 50,000 years to cool from red hot to the temperature of air. Buffon set out in the coming years to conduct his own experiments on the rates of cooling for various substances that make up the Earth. He published his results in a book on minerals in 1774 in which he inferred—he thought conservatively—that it had taken almost 75,000 years for the Earth to cool to its present temperature and over 33,000 years to cool to the point when organic life could begin.

Four years later, in 1778, he brought his conclusions to widespread attention in a book that came to be regarded as a classic of French prose, the *Epochs of Creation*. By this time the so-called Age of Enlightenment, marked by the appearance of numerous ideas that challenged tradition, was in full sway (see Chapter 14). The threat of censure was no longer as immediate as it had been at mid-century, nor was Buffon the only one considering a prolonged age for the cosmos and the Earth. For all that, it was Buffon's *Epochs* that more than any other work introduced to the reading public the notion of an extended period of history, prior to human history, in which no life had existed. That is not to say that he was generally persuasive. Most people in Europe remained convinced that history coincided with human history.

Buffon repeated his theory of the comet striking the sun, splitting off matter that congealed into hot fluid masses circling the sun. He closed this first epoch of creation, which took 2,936 years, at the point when the cooling process had produced a solid Earth that had lost its incandescence. In the next epoch the cooling Earth continued to contract, producing mountains and subterranean cavities over the next 30,000 years. The cooling now condensed the water vapor of the atmosphere in a third epoch, causing torrential rains that covered all but the highest peaks with water. Over the next 20,000 years the first life, shellfish and plants, emerged and thrived in this primitive universal sea. Volcanic activity marked the fourth epoch, a relatively short time of about 7,000 years, opening routes for the water to recede into the Earth and for the appearance of dry land. Animal and plant life appeared on land near the cooler polar regions during the fifth epoch, which lasted another 5,000 years. During the sixth epoch, which lasted the same length of time as the fifth, life migrated toward the tropical regions and the continents began to separate. Some 70,000 years had passed by the dawn of the seventh and final epoch, in which humans appeared.

A Scottish theory of the Earth.   Seven years after Buffon's *Epochs of Creation* appeared in France, James Hutton (1726–1797) communicated to the Royal Society of Scotland his ideas about the Earth's past. This paper appeared in the Society's *Transactions* in 1788, although a longer two-volume work appeared in 1795 under the title *Theory of the Earth, or an Investigation of the Laws Observable in the Composition, Dissolution and Restoration of Land upon the Globe.* Hutton rented out the farmland he had inherited from his father and was able to retire to Edinburgh, where he compiled his reflections about the cause of the Earth's development.

Hutton's understanding of the Earth began with two assumptions that colored his conclusions. First, he believed that the Earth had evidently been made for humankind. For example, a completely solid Earth would not be habitable because plant life requires soil, and humans depend on plant life. It was not accidental that natural forces had broken up the surface of the hard, solid Earth so that it could support life. The second assumption was that the action of nature's forces was not sudden and dramatic, but, as Hutton put it, "the operations of nature are equable and steady." To wear down rock into soil, which was then carried by moving water to the sea, required an enormous amount of time. Hutton observed that "the course of nature cannot be limited by time."

Hutton asked what agency had caused strata of rock, especially those that appeared porous, to form at the bottom of the sea. After much consideration he rejected the possibility that they had been precipitated out of the water, embracing rather the idea that they could only have been fused by the action of subterranean heat. Hutton next asked how materials that collected at the bottom of the sea were raised above its surface and transformed into continents of solid land. Once again he concluded that this had not occurred by the receding waters of the sea, but by the same action that fused the strata—a deep subterranean heat. The continuous action of heat explained how and why the Earth's surface had changed.

The final section of his paper was where Hutton introduced his most unusual claim. He believed that the development of the Earth as he had described it was just a part of a larger, more general process. He thought that, by asking about the state of the Earth *before* the present land appeared above the surface of the waters, he could acquire some knowledge of the larger system that governed the world. He asked, for example, where the materials at the bottom of the sea, fused by heat into the strata he had identified, had come from. Some of the strata were composed of the fused remains of marine organisms. But where had *they* come from?

Hutton could only conclude that there had been a phase of development prior to the one he had described. He even allowed that plant life had flourished on land during this phase; indeed, he asserted that cycles of decay and renovation constituted nature's means of sustaining plant and animal life according to a wisdom that had been in operation for indefinite successions of ages. Hutton may have been convinced that his scheme reinforced the central role life played in nature, but he convinced no more people than Buffon had that the age of the Earth went back indefinitely. Few found any comfort at all in the famous declaration that ended Hutton's treatise: "The result, therefore, of our present enquiry is, that we find no vestige of a beginning,—no prospect of an end."

Hutton differed from Buffon and de Maillet in his concern to demonstrate, through geological phenomena, the presence of a divine plan governing terrestrial processes. Other deists from his time argued that the time scale for geological change could not be constrained by traditional interpretations of Genesis. Hutton agreed, because of the immensely slow pace at which he believed the forces of heat acted. But Hutton differed in his emphasis on the purpose behind the divine plan, one that accommodated the interests of plants and animals. The reason why there were volcanoes was to raise strata so that rivers could break rocks into soil for living organisms. Hutton's work stood as an example of how deism compared with natural theology, in which the knowledge of nature supplied a basis for inferring God's control over nature. For Hutton the prolonged time scale of Earth's past revealed intelligent design.

Mineralogy and Earth History

As noted, the eighteenth-century theories of the Earth, with their emphasis on causal law, stood in a tradition that reached back into the seventeenth century. The search for the causal agency that had molded the Earth according to some grand natural law, however, was not the only means of access to an understanding of the Earth's past. Paralleling these theories of the Earth was another approach, also with roots in the early modern period, in which the starting point was the composition of the Earth's minerals. Minerals traditionally included four classes: "earths" (including rocks), metals, salts, and sulfurs. Scholars gathered information about these various forms of solid materials found on Earth and then, in light of the data assembled, suggested how such diversity might have come about. Here the emphasis was on the description of the development of the Earth from its beginnings to the present in all its complexity, without reference to a single unifying causal law.

The German mineralogical tradition.   This treatment of the history of the Earth was common in the German states, where the presence of rich deposits of ore drew primary attention to metals and accounted for the long-established mining tradition in regions such as the Erz Mountains of Saxony. Mining officials wanted practical information about the location and properties of such valuable metals as lead, copper, and silver. In an earlier period, state-appointed officials drew on individuals trained in the universities to oversee the acquisition of mineralogical knowledge, but in the eighteenth century they established separate technical schools for the purpose of training the officials they required. As a result, the German approach to mineralogy expanded beyond a primary interest in metals.

German mineralogists of the eighteenth century began to subject rocks, previously regarded as mere conglomerations of individual minerals not worthy of study in their own right, to classification. They categorized rocks according to the effect heat had on them (the "dry way"), the purest rock being that whose mineral content remained unaltered even by intense heat. They began to gather more information than just the mineral content of the rocks; for example, they recorded such things as the elevation of rocks, their fossil content, their contour, and they listed impressions about the age and mode of origin of the rocks as well.

For "earths" other than rocks mineralogists preferred the "wet way," which could run in two directions. They could test the solubility in water of minerals with names like magnesia and baryte, or they could precipitate out the mineral contents of waters from hot springs and health spas. Chemists, who were also interested in earths, contributed their understanding of the interactions of earths with acids and bases. From numerous investigations experimenters differentiated a whole range of earths based on their solubility.

Where the history of the Earth was concerned there was widespread acceptance among eighteenth-century mineralogists that the original ocean referred to in Genesis had been a thick gelatinous aqueous fluid made up of minerals in solution, and that rocks and most other solid minerals formed over time by a process of consolidation. There were various explanations of exactly how the consolidation of rocks occurred, but the end result was that rocks began to form and were laid down under the primal ocean. The oldest rocks, which formed the Earth's core, were therefore almost the same age as the Earth itself. On top of these primal rocks came others, whose diverse properties reflected the variations in the contents of the primitive ocean waters at different locations. Eventually the primitive waters diminished, at least in part due to evaporation, and the rocks were exposed to air. The variation in position that rocks now displayed as exposed land was a reflection of the effect that underwater motion— sometimes chaotic and sometimes calm—had on the pattern of their consolidation.

Abraham Werner and the Freiberg School. The most influential figure in the history of geology from the late eighteenth century was Abraham Werner (1749–1817), who studied and later taught at the mining academy in Freiberg. Werner inherited the general ideas described in the preceding section from his predecessors and he openly acknowledged his debt to them. Although he published little, his powerful influence on his contemporaries and on a succeeding generation of scholars on the Continent and in Britain came from his considerable abilities as a teacher and from the new system he created. As a result, his influence extended well beyond Saxony and lasted into the third decade of the next century.

Werner's greatest contribution was to articulate that the period during which rocks were formed, rather than their mineralogy, was their most important feature. He gave to geology the historical entities that he called formations, which were rocks that had been produced in the same period. Werner focused on the variety of information mineralogists had begun to gather about rocks. Unlike his predecessors, who relied only on mineral content when classifying rocks, his goal was to develop a systematic knowledge of all the data gathered about individual regions in order to determine when and how their rocks had been laid down. He called his new approach geognosy, based on the Greek word for abstract knowledge, to emphasize the intellectual reasoning needed to put together the results of careful and widespread observation.

In Werner's account of the Earth's history, the oldest rocks from the calm waters of the primeval ocean, many of which were crystalline but also included metals, consolidated in successive individual formations to form a "primitive class." Next came a small class of formations he called transition rocks, some of which had formed in turbulent waters. The third class of formations he called stratified rocks; some of these resulted from mechanical pressure while others consolidated by chemical means. The final class of formations, called the recent class, came from eroded material deposited by moving water and from the extruded material of volcanoes.

Contrary to a widespread impression, Werner did not appeal to sudden and dramatic events to explain how the Earth had developed; rather, like Hutton, he held that the processes going on in the present were the same as those in the past. For example, he believed that the primeval ocean had gradually retreated over time and that there was evidence to indicate that the retreat had occasionally reversed itself. Werner preferred not to endorse speculations about where the retreating water had gone. He believed it was sufficiently clear that the waters *had* retreated and that speculating about the cause was relatively unimportant. However, late in his life he invoked the new knowledge that water was composed of gases to suggest that primal waters had decomposed when forming the atmosphere.

At the close of the eighteenth century Werner joined others who were willing to extend the history of the Earth far beyond the six thousand years inferred from a literal reading of the Old Testament. Once again, his preference was not to speculate about matters that did not easily lend themselves to precise determination. The most he would concede was an oblique reference to a time "when the waters, perhaps a million years ago, completely covered the earth." As the century came to a close, then, there was ample indication that natural philosophers had begun to accept the Earth as a cosmic body whose past had been shaped by natural processes that they were responsible to identify and comprehend.

Suggestions for Reading

Michael Crowe, *The Extraterrestrial Life Debate, 1750–1900* (New York: Dover Publications, 1999).

Thomas Hankins, *Science and the Enlightenment* (Cambridge: Cambridge University Press, 1985).

Rachel Laudan, *From Mineralology to Geology* (Chicago: University of Chicago Press, 1994).

Mary Terrall, *The Man Who Flattened the Earth: Maupertuis and the Sciences in the Enlightenment* (Chicago: University of Chicago Press, 2006).

Photo Credits

CHAPTER 1
2 Mary Evans Picture Library/The Image Works.

CHAPTER 2
31 © Musée Condé, Chantilly, France/The Bridgeman Art Library; **32** From *Que hoc volumine contine (n) (tur:) Liber de intellect; Liber de sensu; Liber de nichilo; Ars oppositorum…insup (er) mathematicu (m) opus quadripartite (m): De numeris perfecti.* Parisiis: Ex officina Henrici Stephani, impsesis eiusdem et Ioannis Parui, 1510, primo cal. (1) Februari. Call number 798206. Image, leaf 63. By Courtesy of the Department of Special Collections, Memorial Library, University of Wisconsin–Madison.

CHAPTER 3
55 Bibliothèque Nationale, Paris, France/Giraudon/Art Resource, New York; **59** Time Life Pictures/Getty Images; **63** Mary Evans Picture Library/The Image Works.

CHAPTER 4
72 © Bibliothèque de la Faculté de Médecine, Paris, France/Archives, Charmet/The Bridgeman Art Library;

74 Marine satyr from *Historiae Animalium,* © Academy of Natural Sciences of Philadelphia/Corbis; **79** From *A Book on fossil Objects, chiefly Stones and Gems, their Shapes and Appearances,* by Conrad Gesner, 1565.

CHAPTER 5
100 © Bettmann/Corbis; **106** © Bettmann/Corbis.

CHAPTER 7
142 © Bettmann/Corbis; **145** From *Philosophe Naturalis Principia Mathematica,* by Isaac Newton, 1666; **152** Hulton Archive/Getty Images.

CHAPTER 8
169 *Portrait of Sir Isaac Newton* by Godfrey Kneller, 1689/ © Bettmann/Corbis.

CHAPTER 9
182 ©Bettmann/Corbis; **187** Stocktrek Images/ Royalty-Free/Getty Images.

Index